Fatal Flaws

Fatal Flaws

How a Misfolded Protein
Baffled Scientists and
Changed the Way
We Look at the Brain

Jay Ingram

WITHDRAWN

Yale UNIVERSITY PRESS

New Haven and London

Published with assistance from the Louis Stern Memorial Fund.

First published in Canada in 2012 by HarperCollins Publishers Ltd.
First published in the United States in 2013 by Yale University Press.

Yale University Press books may be purchased in quantity for
educational, business, or promotional use. For information, please e-mail
sales.press@yale.edu (U.S. office) or sales@yaleup.co.uk (U.K. office).

Set in Minion type by Integrated Publishing Solutions, Grand Rapids, Michigan.
Printed in the United States of America.

Library of Congress Cataloging-in-Publication Data
Ingram, Jay.
Fatal flaws: how a misfolded protein baffled scientists and changed the way
we look at the brain / Jay Ingram.
        p.   cm.
Includes bibliographical references and index.
ISBN 978-0-300-18989-6 (hbk. : alk. paper)
1. Prion Diseases—physiopathology.   2. Prion Diseases—history.
3. Prions—history.   4. Prions—pathogenicity.
HV6457.WL301 2013
616.8'3—dc23
2012042874

A catalogue record for this book is available from the British Library.

This paper meets the requirements of ANSI/NISO Z39.48–1992
(Permanence of Paper).

10 9 8 7 6 5 4 3 2 1

To M. A.

Contents

Fatal Flaws

Introduction

It isn't an image that immediately grabs you: just a bluish background peppered with irregular white shapes. If it weren't for the colors, it could be a high-resolution aerial view of the Canadian Shield, myriad lakes dotting an endlessly rocky landscape. There is one other feature: thousands upon thousands of tiny rust-colored dots. A scale bar, measured in micrometers, tells you you're looking at a photomicrograph, but of what?

It isn't an image that immediately grabs you, unless you know *what it is:* an extremely thin slice through the brain of the first cow in Canada ever to be diagnosed with BSE—mad cow disease. I am studying this image in a microscope in Dr. Stefanie Czub's lab at the Canadian Food Inspection Agency in Lethbridge, Alberta.

Imagine a razor-sharp knife cutting through a fruitcake. Cherries could be sliced lengthwise, across or at an angle, looking different every time. That's the case here: the blue background is a welter of neurons, packed together, each one sliced through at a different place. So some show up as round and full—that's when the microtome knife hit the neuron at a right angle. But in other places the bluish cellular material is long and thin, where the blade skimmed along one of the neuronal extensions that reach out to other cells.

Those "lakes," the white areas that look like holes? They *are* holes—vacuoles—some natural and to be expected, but most of them obvious signs of disease.

And the rusty dots? Deposits of prions, the mysterious infectious agents responsible for mad cow and other diseases, agents that we are only now beginning to understand. There are untold numbers of them in this tiny slice of the unfortunate cow's brain, and they caused her death.

It's one of those moments when the object of your attention (in this case a minute portion of a single microscope slide) is so compelling, you completely forget the setting around you. But it is nonetheless a remarkable scene. Here I am, in a building on the banks of the Oldman River, in the wild and windy foothills of southern Alberta, dressed in the typical containment lab outfit of disposable suit, shoe covers and gloves, staring at a microscope slide absolutely loaded with deadly infectious agents. Deadly still, even though they have been stored inert on glass for nine years. That incredible persistence is just one of the puzzling qualities of prions, agents that had, in causing BSE, created an agricultural and public health disaster in the United Kingdom, and which are so baffling, so out-of-the-ordinary, that they forced the creation of a new science devoted to understanding them.

This is an account of the birth of that science. And what a science. It appeared out of nowhere and not, as is usually the case, in centuries past, but mere decades ago. It is still changing before our eyes, making this a unique opportunity to watch how a science develops. And more than that, it is a science that hits at the heart of the food we eat and many of the illnesses we die from. And because it is based on an absolutely heretical idea that mere molecules (prions) can infect and kill a wide range of animals, it challenges the very foundations of biology. Never before had scientists discovered a disease-causing entity that has no genes. Even viruses, as stripped-down as they are (so much so, they are incapable of reproducing on their own), have genes that allow them to reprogram the cells they have infected. But prions don't even have the skimpy complement of genes a virus has. They are just protein molecules with no genetic material at all, yet they can infect, multiply and kill. Scientists had never even *envisioned* something like that. Indeed, there is still a handful of scientists who don't buy it, even after decades of research.

So the arrival of prions on the scientific scene has been dramatic, but not just for the fact that this field has broken new ground. Along the way, the struggle to establish radically new ideas has displayed—vividly—what you might call the "seamier" side of science, features of which the public is generally unaware: the ambition, competitiveness, bitterness, flights of creative brilliance, character flaws and human foibles of researchers; the tendency of those outside science, especially governments, to ignore or suppress research that doesn't tell them what they want to hear; even the reluctance of science itself to accommodate itself to new ideas.

Why do all these normal attributes of science almost always fly under the popular radar? Largely because we cling to the same tired old images of scientists today that we always have had: we imagine them as emotionless, nerdy, unfailingly rational, inadequately socialized people, often clothed in lab coats, lost in thought while peering into a microscope. They seem not to be human like the rest of us. But, of course, they are.

The scientist who stands out from all others in the world of prions is Stanley Prusiner, Nobel Prize winner and implacable foe to those who do not share his views. But Prusiner is just one in a long line of scientists who have the same attitude to "competitors." Take the inhumanly brilliant Isaac Newton: he fought with Gottfried Wilhelm Leibniz over the credit for inventing calculus, with Christiaan Huygens over the wave theory of light and with fellow countryman Robert Hooke over any number of matters of priority. My favorite example: Newton is often quoted as having admitted that "if I have seen further, it is by standing on the shoulders of giants."[1] Most approve of this as an admission by the great man that he couldn't have done it all alone. But others, noting that Hooke was reputedly a hunchback ("crooked" is the word used by his contemporaries), see malevolence in those apparently gentle words.[2]

But science is conflicted in many ways, not just by feuds between individuals. Most research is funded by governments in one way or another, and while you might suppose that such funding should be arm's-length, with no strings attached, allowing the scientists to do what they do best, in fact, governments are usually very busy ensuring that scientific messages that *they like* are the ones getting out to the public.

Prion science provides a beautiful example. Because the most controversial and devastating prion disease, bovine spongiform encephalopathy (mad cow, or BSE), was an immediate and urgent threat to British meat exports, government officials suppressed the science, fearing a body blow to the

British economy. But they weren't doing anything that their colleagues past and present wouldn't have done. There is a colorful history of governments contradicting, ignoring or suppressing science.

One example of many: as the automobile industry in the United States was taking off in the 1920s, General Motors was gaining ground on Ford and its beloved Model T, but it needed something to add to gasoline to eliminate engine knock. GM scientists, led by Thomas Midgley, discovered that the compound tetraethyl lead, when added to gasoline, provided that extra kick. But public health experts immediately protested that lead had been unambiguously shown to be toxic to the human nervous system. GM nevertheless asserted that, *with proper care,* lead was safe and convinced the government of the day to put its stamp of approval on it. The safety claim was powerfully demonstrated by Midgley, who publicly washed his hands in leaded gasoline and dried them on his handkerchief, then calmly inhaled the fumes for a full minute—this after he had earlier been forced to take a leave of absence from work to recover from lead poisoning. It was another sixty years before leaded gas was banned—because of its effects on health.

Midgley's piece of theater eerily foreshadowed a similar gesture at the height of the mad cow disaster. In 1990, then British minister of agriculture John Gummer made a public show of not only eating a hamburger made from British beef but also encouraging his somewhat reluctant four-year-old daughter, Cordelia, to do the same. Six years later, young Britons were dying grotesque deaths from having done essentially the same thing.

And even when governments or businesses have no particular axe to grind with science, science itself may stand in the way of change. Not very long ago, stomach ulcers were assumed to be the result of excess stomach acid brought on by stress, smoking or bad genes. But Australian microbiologist Barry Marshall became convinced that a bacterium, *Helicobacter pylori,* was the cause. Desperate to convince an extremely skeptical medical community, which refused to believe that any bacteria could survive in the acidic environment of the human stomach, Marshall was driven to perform the near-heroic act of swallowing a rich culture of the bacteria to prove his point. Within days he was ill, reinforcing the idea that these bacteria were indeed pathogenic and providing at least indirect evidence (given the gastric distress that Marshall was suffering) that they might contribute to the formation of ulcers.[3]

In Marshall's Nobel Prize acceptance speech he remarked that the belief that ulcers were stress- and cigarette-related was "akin to a religion" and that the extreme skepticism that greeted him was due to the belief that the existing ulcer treatments (drugs and surgery) were the best medicine could do.

As the science of prions developed in the 1980s and 1990s, there was similarly powerful skepticism of the idea that a microscopic agent could lack genes but still be infectious. In the case of prions, however, the antagonism wasn't directed just at the idea but also at the man who was its most aggressive proponent, Stanley Prusiner. To say that Marshall and Prusiner reacted to their skeptics differently is putting it mildly: Marshall admits to being impatient with road-blocking colleagues but mostly kept his comments to himself. Prusiner expressed his impatience (even disdain) publicly, if thinly cloaked in the language of scientific journals.

There is another, more essential, difference in the stories. Once it was established that peptic and duodenal ulcers are caused by a bacterium, and that it was sensitive to antibiotics and therefore curable, all that remained to be done was to work out the details. But prions are different. There is so much about them that is still unknown. How exactly do they propagate? What makes them invariably fatal? And probably the most important question of all: are we looking here at a fundamental feature of all the important chronic neurological diseases, something we never recognized about them before?

The prion story is far from over; it is a rough-and-tumble affair, with rivals, eccentrics, interfering governments and brilliantly creative people all getting in a lick now and then. And we are lucky enough to have a ringside seat for it all.

This story begins in Chapters 1, 2 and 3 with the discovery by Western scientists in the 1950s of kuru, a completely unheard-of disease unique to a small territory in Papua New Guinea. Kuru was a sensation, attracting tabloid readers and medical scientists alike, and no wonder: it was exotic, scientifically baffling and invariably fatal, and carried with it whispers of cannibalism.

Kuru had something for everyone, and it lit the fuse: soon researchers were criss-crossing the New Guinean jungle, taking blood samples, comparing symptoms and trying—unsuccessfully—to help stricken patients. As told in Chapter 4, two Americans, one in England and one in Washington, DC, then surprised everyone by showing how kuru is part of a family of diseases, the others more familiar but still poorly understood.

Prion science is a child of the molecular biology revolution begun by Watson and Crick in 1953. Prion diseases destroy cells by infecting them in a way that scientists not only had never seen but could never have anticipated. The subtlety of this attack can be appreciated only by looking at life and death at the *molecular* level, and Chapters 5 and 6 provide a glimpse of that.

With that background in place, we return in Chapters 7 and 8 to the story, now running full speed on two separate tracks. On one, scientists in England in the 1960s wrestle with the peculiar nature of one of the kuru-related diseases, the affliction of sheep called scrapie. Although the disease had been known for centuries, the nature of the agent causing it was turning out to be so strange that most biologists simply couldn't accept the evidence coming out of the lab. It wasn't that they hadn't thought about something like this; it was that they *couldn't*. It was biological heresy, and was described in exactly those terms.

The other track was disturbing in a different sense. Through the 1970s, scientists around the world began to realize that the agents of Creutzfeldt-Jakob disease (CJD), the other human disease of the prion kind, were practically immortal, and could spread and kill even in environments that sterilized all other known pathogens.

In the end, the two tracks would converge, but no one knew that at the time. At this point in the story, once again, it's all about the molecules, but specifically the single most important components of life, the proteins. Chapters 9 and 10 show how these are, with a tip of the hat to DNA, the true molecules of life. Proteins are responsible for building, maintaining and running the living cell, but, as scientists were beginning to suspect, protein molecules are also capable of ending life.

Then, in the 1980s, all hell broke loose. As described in Chapters 11 and 12, Stanley Prusiner infuriated and alienated colleagues (while impressing countless others) by proclaiming that he (and, implicitly, he *alone*) understood the infectious molecule that was confounding them all. He had the gall to name these molecules "prions," a word of his own invention and which implied a claim of ownership. Eventually, Prusiner was awarded a Nobel Prize for his insights, showing that in the end it's what you discover, and not so much how you discover it, that counts—at least with Stockholm. Undeniably, however, the picture being painted of prions was stunning, an opening into a world of infection that no one had dreamt of—that, in Chapter 13.

Then, in what surely has to be one of the most sensational coincidences of all time, at the height of the scientific clash, with researchers lining up for or against Prusiner, the United Kingdom was brought to its knees by a whole new and utterly devastating disease of this very kind: mad cow disease. Chapter 14 chronicles this part of the story, highlighted by the struggle between managing unsettling information and allowing science to be public.

As if it weren't enough that the story, even to this point, was a complicated mix of biology, medicine, human tragedy, disappointment and disbelief, it intensified with the horrifying dis-

covery in the mid-1990s that people who had eaten beef products during the height of the mad cow epidemic—even very
young people—were now themselves susceptible to a human
version of the disease, characterized by a protracted decline
into dementia and death. That story is told in Chapter 15. And
while the vast majority of those who have succumbed so far
have been residents of the United Kingdom and France, we in
North America have our own worries.

Although we can be thankful that apparently we have had
no domestic cases of the new human disease, variant CJD, we
have certainly been clobbered by the discovery of mad cow
disease in Canada. So far, eighteen cows have been diagnosed
with the disease since 2003, and the Canadian beef industry
has taken the hit, losing billions of dollars of beef exports. The
economic and social costs of a disease that has not struck one
Canadian have been enormous, and the science, true to form,
puzzling.

As we see in Chapter 16, there was evidence decades ago
that there was a prion disease in North American cows, but
that evidence was hurriedly dismissed by officials in the United
States, a country that somehow, by some strange quirk of fate,
even though its national cattle herd is at least five times bigger than the Canadian, and even though Americans fed their
cattle more meat and bonemeal than Canadians did, has had a
mere two cases of BSE, abnormal BSE at that.

But these days, concern over the North American cattle
herd is taking second place to worries—even fears—of another
prion illness, chronic wasting disease (CWD). As described in
Chapter 17, it first appeared in captive deer in Colorado in the

late 1960s but soon spread to wild deer, then elk and moose, then to other states and provinces. It continues to spread today. Like all prion diseases, it invariably kills the animal it has infected. How is it spreading? No one knows. How far might it spread? Again, no one is sure. What other animals might be susceptible to it? That is the million-dollar question.

Consider the vast migrating herds of caribou in northern Canada. Infected deer are already known to be moving farther and farther north in Saskatchewan. If those deer meet the caribou, and if they are leaving CWD prions behind them, and if caribou are susceptible . . . then we have the potential for a disaster, both for the caribou and for the people who depend on them. Might humans even be susceptible to CWD? Despite a few suspicious cases, the jury is still out on that question. But even if we are not, some experts fear the devastation of cervid populations (deer, elk, caribou and their relatives) across Canada over the next century.

In the face of that pressing issue, we find out in Chapters 18 and 19 that the science of prions is still in its infancy. In the labs today, scientists are still struggling with fundamental questions: How do infective prions get from one animal to another? And where and how did these diseases start? Because they are caused by the misfolding of normal proteins, it would be crucial to figure out what the normal version does in our cells, particularly in our brains. But even now, no one knows the answer to that.

But beyond the molecular biology or, rather, as a part of it, there are now unsettling questions about the relationship between Creutzfeldt-Jakob disease and other, much more common, neurodegenerative diseases, like Alzheimer's disease, Parkinson's disease and amyotrophic lateral sclerosis, or Lou Gehrig's disease. Ten years ago, a claim of a connection among these would have been dismissed by all but a very few, but now scientists know that at the molecular level there is some sort of link, some kind of common mechanism, even though at this point no one is suggesting that these other diseases are infectious. Most recently, chronic traumatic encephalopathy, the brain disease associated with repeated blows to the head, has exhibited its own prion-like qualities. Chapters 20 to 23 show just how tricky clarifying such relationships among these diseases can be.

In the end, the science of prions has been a scientific, political and social roller coaster, and we have not seen the end of it yet. It has also laid bare the intensely personal side of science, as portrayed in Chapter 24. While the future of the research can't be clearly made out, one thing is certain: the past fifty years of this science have provided some of history's most stunning revelations about disease, the brain and infection.

1

The Mystery of Kuru
A Disease Like No Other

Even as the world succumbs to globalization, there remains a handful of places that have resisted exploration and continue to offer the opportunity for discovery. The island of New Guinea is among the most spectacular. Researchers are still able to find one more valley there, another mountaintop, where they can encounter never-before-seen species of birds, mammals, amphibians and invertebrates. It is the physical geography that makes it so. New Guinea is so big it practically qualifies as a continent—its just over 300,000 square miles make it a little bigger than the province of Alberta—but it retains the exotic biology of an island, separate from the tides of evolution that sweep continents. And unlike Canada, with its unending stretches of shield, forest, tundra and prairie, the only thing unending about New Guinea is change and novelty. Rugged beyond imagining, every new valley, river or mountainside provides a barrier to migration; each ecological niche sequesters the organisms in it.

This patchwork terrain confines its human inhabitants as well. The island is home to nearly a thousand languages, out of the world total of somewhere between six and seven thousand. This in a population of a mere seven and a half million. Languages, like species, differentiate and preserve their differences wherever geographic barriers minimize contact.

It's still possible in the twenty-first century in New Guinea to climb a peak and report—for the first time—the mating ritual of Berlepsch's six-wired bird of paradise (a bird known only by its remains in museums), to discover birds that use the same poisonous chemicals in their own defense as poison dart frogs do half a world away, to find dozens of new species of frogs and even a giant rhododendron, many of these in settings that even the locals have never set foot in. Now imagine what it was like at the end of World War II.

After Japan surrendered, the Australian government took control of the eastern half of the island, now called Papua New Guinea, and through the late 1940s and early 1950s sent administrative and medical personnel to the island. Almost immediately, reports began to drift back of something very strange occurring in the jungle: a disease unlike any seen before. Called kuru, it affected only one well-defined group, the Fore people of Papua New Guinea's Eastern Highlands, and it was invariably fatal.

When Westerners first appeared on the scene, most of the Fore had never even seen the oceans that surrounded them on all sides, although they had been rudely introduced to Western culture by the crash of the odd World War II fighter plane a few years before. They lived mostly in small villages, with land cleared around them for the cultivation of crops like sweet potatoes. They raised pigs for protein. There was nothing about the Fore that would suggest they were in any way unusually medically compromised, although, in the absence of antibiotics, yaws, a flesh-eating disease caused by a relative of the syphilis organism, was allowed to progress to disfigurement. But there is nothing mysterious about yaws. Kuru was different.

Anthropologists first recorded it in 1951 and 1952. A typical report was one from 1953 in which the witness, a government patrol officer, described a young girl sitting by a campfire, shaking and jerking. He was told that she was the victim of sorcery and would be dead within weeks.

Two years later, the first Western medical description of kuru was recorded, but its label of "acute hysteria" wasn't much more helpful than "sorcery." Nonetheless, kuru had now become more than simply the stuff of third- or fourth-hand rumors. There was a disease in New Guinea, it was fatal and, more intriguing to scientists, it could not be explained.

The first symptoms were an unsteady gait and slight decrements in coordination, but these signs worsened rapidly, so that unsteadiness and uncertainty gave way to the inability to walk unaided and, finally, to stand up. No one survived kuru. These symptoms strongly suggested that it was a brain disease, but through most of the steady decline the afflicted person was mentally intact, suggesting that if the brain was involved, it might only be a part of it, perhaps the cerebellum, the bun of tissue at the back of the brain that coordinates movement. Although anthropologists were the first to record it, it wasn't long before kuru (a word meaning "trembling" or "fear" in the Fore language) began to attract the attention of dynamic and well-connected medical detectives, the most industrious and charismatic of them Carleton Gajdusek.

Gajdusek's life was a complex and controversial one (he is

the only Nobel Prize winner to have been convicted of pedo-
philia and imprisoned), but there's no doubt that he had an
enormous impact, not just on the study of kuru but also on the
entire branch of medicine that has evolved from it. Gajdusek
had enormous physical and mental energy, and he brought
both to bear on the mystery of kuru.

The vast number of letters Gajdusek wrote from New
Guinea to an array of medical associates provides a glimpse of
how remarkable the man was. And in his Nobel Prize address,
he referred to *five thousand pages* of his own journals on his
"explorations and expeditions to primitive cultures." When the
man found time to write so voluminously is a mystery. But his
ability to set a scene was impressive. Upon his arrival in New
Guinea, he wrote: "I am in one of the most remote, recently
opened regions of New Guinea (in the Eastern Highlands) in
the center of tribal groups of cannibals only contacted in the
last ten years and controlled for five years—still spearing each
other as of a few days ago, and only a few weeks ago cooking
and feeding the children the body of a kuru case, the disease I
am studying."[1]

Disease, murder and *cannibalism!* Despite the ominous tone
of this description, the people Gajdusek encountered were
generally friendly, but that hardly made his work any easier.
Here's an excerpt from his detailed description of a thousand-
mile hike he took to map out the furthest reaches of the disease: "The two days' dense jungle trip from isolated Kasarai to
the Yar people has now become three days. The track was not
only impossible to find, but impossible to walk without a crew
in front to cut, bridge and bushwhack the way that old Anuma

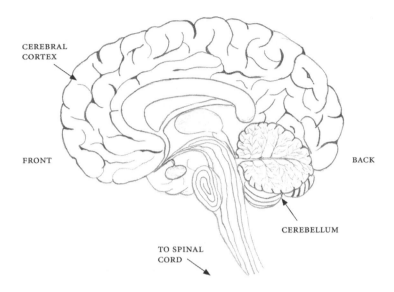

CEREBRAL
CORTEX

FRONT   BACK

CEREBELLUM

TO SPINAL
CORD

The right half of a normal brain that has been sliced down the middle. Ancient brain structures are clustered together in the center with the cerebral cortex (the "thinking" brain) above. Lower right is the cerebellum, which is hit hardest in kuru.

indicated. For the life of us, we could not see how he kept his bearings; but we came through, dropping to well below 3000 feet into real tropical jungle at the rushing river that goes from Urai village down to the Yani. This river we had to ford thrice, and once crossed precariously on an immense tree-bridge."[2]

And if that weren't impressive enough, the next day he adds: "Mosquitoes and leeches are terrible, and huge insects of every description abound. I will make a collection of those only over six inches long for photography tomorrow."[3]

These adventures were much more than hiking and climbing: at the same time, Gajdusek was collecting urine and blood samples, performing autopsies, preserving tissues for shipping to Australia and the United States (including brains), all the while racking his brain for an explanation for the disease.

Yet, as impressive as he was, Gajdusek rubbed many people the wrong way. Sir Macfarlane Burnet—an Australian Nobel Prize winner who resented that Gajdusek, an American, was leading the charge against kuru, a disease in an Australian-administered territory—said he'd been told that "Gajdusek was very bright but you never knew when he would leave off work for a week to study Hegel or a month to go off to work with Hopi Indians." He'd also heard that "the only way to handle him was to kick him in the tail, hard. Somebody else told me he was fine but there just wasn't anything human about him. . . . My own summing up was that he had an intelligence quotient up in the 180s and the emotional immaturity of a 15-year-old. . . . He is completely self-centred, thick-skinned and inconsiderate, but equally won't let danger, physical difficulty or other people's feelings interfere with what he wants to do."[4]

A near-genius with a swashbuckling temperament, as one described him. Indiana Jones with extra empathy. But Gajdusek didn't care what other scientists thought; he had developed strong friendships with the Fore people, he recognized the intellectual challenge of kuru and he wanted to be the one to overcome it.

In the beginning, Gajdusek had no way of knowing exactly what was going on inside the brains of kuru victims—until he gained the confidence of the Fore, he would not be able to get his hands on the brain tissue of any of those who died of the disease. And in the absence of hard evidence like that, there wasn't much to go on. The Fore themselves (who were better at recognizing the very early signs of the disease than Gajdusek) had told anthropologists that kuru was a relatively recent disease, dating back only to the 1920s or so in the north, and even more recently in the south. At first they themselves had mistakenly thought that the condition was temporary, perhaps caused by spirits of some kind, but when it grew clear that it was invariably fatal, they became convinced that the disease was the result of evil sorcery. As a result of this suspicion, it wasn't just kuru itself that was fatal—accused sorcerers were attacked and killed as well.

By the time Gajdusek and others had started their investigations, the disease had become a threat to the very existence of the Fore: as many as two hundred people were dying of it every year in the late 1950s. Its impact was exaggerated by the fact that it was selective in its targets. Women and children of both sexes were vulnerable, adult men much less so. As a result, widowed men and motherless children were becoming more and more common.

The cause was not at all obvious, and Gajdusek and those who worked with him had to throw their diagnostic net as wide as possible. Could it be some toxic contaminant in the food, or even in the soil around the villages? Determining if that might be the case seemed hopeless—there were just too many potential contaminants to consider, and the lab technology necessary was too painstaking, tedious and, as a result, costly. This was, after all, deep in the New Guinean jungle. Gajdusek wrote: "No one will appreciate having to study toxicologically the 500-odd foodstuffs eaten by our Fore kuru patients (as well as all other natives hereabouts); and thus far we can find nothing peculiar to either kuru patients, kuru-affected populations, households with currently active kuru, or the age- and sex-group predominantly suffering from kuru. Without such a 'lead' we are faced with the impossible task of toxicologically surveying some 500 species and varieties of animal and vegetable life of which the Fore partake."[5]

If the men were eating away from the villages and/or if some food that most women and children ate could become toxic upon storage, the gender and age imbalance might make sense. There was, in the medical literature, a precedent: the vitamin deficiency pellagra, which had been common in the southern United States. It too exhibited an asymmetry between men and women. One investigator wrote of kuru: "The data on age and sex reveals a selectivity for women and children. I am not acquainted with any such sex ratio in a hereditary disorder, but am reminded of a similar age and sex incidence in the careful surveys of Goldberger and associates on pellagra. The possibility that adult males might be eating more often away from home would suggest either a deficiency disease or, more likely, that some food substance—perhaps one that is relatively new to this population, or one that becomes toxic in storage and which women and children are more likely to eat—could account for the report in age and sex distribution."[6]

Superficially similar maybe, but in the end, pellagra and kuru had nothing else in common.

If it wasn't caused by a food substance (and there wasn't really any evidence that it was) or by a nutritional deficiency, could there be some sort of environmental contaminant that gained access to the patients' bodies in some other way? Smoke was a possibility: men and women lived apart, the men in large communal houses, the women and children in individual huts. Those huts were constantly filled with the smoke of cooking fires, which in such concentrations could conceivably be a source of excess carbon monoxide, a known cause of neurological damage. And then there was one of the most obvious possibilities: infection. Tropical rain forests are paradise on earth not just for birds and animals but also for parasites, fungi, bacteria and likely viruses too.

That might have been an obvious possibility, but one of the most puzzling attributes of kuru was the complete absence of any signs of infection: no fever, no inflammation, nothing accumulating in the blood or spinal fluid, none of the classic signs that the body's immune system had geared up to fight off something that it recognized as foreign. Even in the absence of these reliable markers, some experts still clung to the idea of an infection, but not Gajdusek. To him it was nothing more than a wild possibility. As far as he was concerned, kuru patients simply didn't look infected, and after all, he was on the ground with them and was the one with the most profound knowledge of the disease.

Some experts thought the Fore had to be carrying some gene that predisposed them to kuru, a lethal gene that selected its victims by age and gender, affecting women and children more than men. In some ways it seemed reasonable. Kuru ran in families, and it exhibited the phenomenon known as "anticipation," where each generation became afflicted earlier in life than the previous one, a generational acceleration typical of some genetic conditions. It was also true that those who emigrated from the Fore territory to an area in which kuru was rare sometimes came down with the disease anyway, whereas those who immigrated into the danger zone usually remained untouched. Gajdusek, for one, thought that the disease was likely a combination of genes and some hard-to-identify environmental factor.

The genetic hypothesis was tricky, though: the disease-causing gene would somehow have to affect women four times as often as men. One such scheme suggested that a kuru gene would act differently in the two sexes, dominant in women (even one copy of the gene would be lethal) but recessive in men (who would have to inherit a copy from each parent to become ill). It looked reasonable at first: women who inherited two copies of the gene would not only get kuru, they'd get the most serious early-onset form. This seemed at first to explain anticipation, where the disease hit the offspring of kuru mothers earlier than it had struck their parents.

But the genetic math simply didn't work out: for such a gene to be as common as it apparently was among the Fore, it would have to have been favored by evolution in some way, perhaps beneficial in a single dose but fatal in a double dose, in the same way that the gene that causes sickle-cell anemia protects those carrying a single copy against malaria but severely compromises the life of those unlucky enough to receive a double dose, one from each parent. There was no indication of such a benefit for a hypothetical kuru gene. Besides, from all accounts, kuru was a new disease, appearing for the first time in the decades before World War II, and that wasn't enough time for such genes to become established. In the end, even though some believed, like Australia's Macfarlane Burnet, that "the disease will turn out to be almost, or wholly, of genetic nature," the genetic hypothesis never really gained any traction.[7]

These first few years of kuru investigation blazed all kinds of trails, but almost all were dead ends. As dynamic as Carleton Gajdusek was, he couldn't solve the mystery on his own, and in the end, the solution to the puzzle of kuru was stranger than all the speculation.

2

Barflies and Flatworms
How Speculation and Pure
Chance Advance a New Science

As the scientists, Carleton Gajdusek most prominent among them, worked tirelessly collecting blood, hiking hundreds of miles through the New Guinean jungle and interviewing the Fore people about their dietary and cultural customs, they became increasingly frustrated by kuru: none of the medical explanations they were trained to recognize seemed to apply to this unsettling disease.

At the same time, Western old-timers in the bars in New Guinean towns, unencumbered by any medical training, had no doubt what was going on with kuru: *cannibalism.*

No one who spent any length of time in the Eastern Highlands in the 1950s doubted that there was cannibalism going on up there in the hills, but how seriously can the testimony of a guy with a few beers in him be taken at a time when the disease is defying some very astute medical minds?

Except it wasn't just tipplers who were talking. Anthropologists had been in the area since the early 1950s and had reported Fore cannibalism. The Fore version was not a bloodthirsty devouring of enemies, but rather a mortuary feast honoring a family member who had died. These were attended largely by the women and children in a village. There are differing opinions as to the men's participation, some arguing that they ate only muscle tissue, some that they didn't participate at all. Anthropologists had documented the ceremony in some detail: how the hands and feet were cut off, then the muscles stripped from arms and legs, the bitter-tasting gall bladder carefully cut away from the liver, and even how tissues like brain were stuffed into bamboo tubes and steamed before being eaten. These feasts were sloppy affairs, with bits of bone, brain and flesh all over the place. There were vivid descriptions of how celebrants had flesh smeared over themselves and, apparently, weren't going to be washing anytime soon.[1] But belying that image, the feasts were highly organized affairs. The Fore had well-established rules for who got to eat what. For instance, if a man or boy died, his brain belonged to his sister; a woman's brain belonged to her son's or brother's wife.

Even with knowledge of all this, investigators were skeptical that cannibalism might be spreading kuru. For one thing, the Fore apparently weren't the only cannibals in New Guinea, yet they were definitely the only group succumbing to kuru. So what fatal flaw set them apart? More important, for cannibalism to be accepted as playing the key role in the spread of kuru, there had to be something for it to spread, an infectious agent of some kind. As long as potential causes like genetics or smoke or toxic soil chemicals were still in the running, cannibalism didn't seem relevant. Remember, one of the most significant and puzzling aspects of kuru was the absence of typical signs of infection: no fever, no inflammation—features almost always found in infectious disease. This seemed to be no infection. Gajdusek was dubious about cannibalism for this very reason.

"Finally, I can still see no sign of infection or post-infectious phenomenon," he wrote to a contact back in Washington, "but with something as unique and unprecedented as kuru, a unique and unprecedented explanation is required. . . . Cannibalism as a possible source seems well ruled out."[2]

But even with that strong statement, Gajdusek didn't abandon the idea of cannibalism completely. He and others had concocted the idea that possibly infants, having consumed even small amounts of human brain tissue, could become sensitized (in the allergic sense of the word) to it, then experience a severe reaction if reexposed at a later date. He was apparently fond of this concept, once referring to it as a "romantic" one that led him almost to wish that cannibalism was more common, but the theory never went anywhere.[3]

So Gajdusek dismissed cannibalism for the good reason that he wasn't able to make it make sense. It likely didn't help that there were no reports of the transmission of disease by cannibalism from any other culture. There's a good reason for that: it doesn't seem to happen. It is very difficult for cannibalism to be the main route of disease spread for *any* species. Unless there are repeated incidents where several healthy individuals eat an infected one (as in kuru), the disease won't go anywhere. If it's simply one on one, à la Hannibal Lecter, a disease can never affect more than one individual (although in that particular case, even one would be enough if it put an end to the Hannibal series of movies).[4]

But by the early 1960s, momentum was building behind the idea that cannibalism had *something* to do with kuru. And a beautiful coincidence fueled that momentum. As Australian administrators had moved into New Guinea after World War II, they promoted social change, and one of the strongly recommended changes was the cessation of cannibalism. Apparently, the Fore abandoned the practice without much objection sometime in the late 1950s, certainly by 1960, and there was an immediate impact on kuru: the number of young children coming down with the disease began to drop. This decline in the incidence of the disease really made sense only if an infection were somehow involved, notwithstanding the evidence against one. If it were an infection, then children "born after the ban" would have had no opportunity to come into contact with the kuru agent, and so were safe.[5] It was apparently a different story for adults—it soon became clear that even if someone had not eaten or come into contact with infected human flesh for three, four, five years or even more, they could still contract kuru. The incubation period of the disease, the time between the disease setting up housekeeping and the subsequent appearance of symptoms, could apparently be very prolonged. This particular feature of the disease, while anything but new, continues to affect the Fore people even today and has turned out to have troubling implications in other parts of the world as well. More on this later.

But even if the cessation of cannibalism had changed the dynamics of the disease, there was still no definitive link. Nor was it sufficient for someone merely to make the connection (the regulars at the pub had already done that); they had to make the connection *and supply the reasoning behind it*. In the end, that someone wasn't a medical expert at all.

According to anthropologist Shirley Lindenbaum, it was her then husband, Robert Glasse, who stumbled on the possibility of a connection. As she related the story to anthropologist Hank Nelson, Robert was reading the May 18, 1962, issue of *Time* magazine one day when he came across an article about the controversial—but fascinating—experiments of James Mc-Connell at the University of Michigan. Nelson writes, "In it there was a report that if you trained a flatworm to respond to an electric flash, then chopped it up and fed the pieces to another flatworm, the cannibal flatworm acquired the memory of the trained flatworm. Bob asked, how would that be for a model of kuru? And Shirley remembers, 'we laughed ourselves silly.' But over the next few days various strands of knowledge began to weave into one."[6]

If this is indeed the way it happened, the story is stranger even than it seems. That issue of *Time* does have an account of McConnell's work, especially those experiments where flatworm cannibalism seemed to transfer learning from those that had been trained to those that hadn't. McConnell's research was extremely controversial, partly because other labs had trouble replicating his results and partly because the results were, on the surface, so hard to believe. At the time—and even now—learning was thought to result from the strengthening of connections between neurons in the brain, not by the creation of some sort of molecule specific to a memory. It wasn't long after the *Time* article, and certainly by the mid-1960s, that McConnell's research was no longer taken seriously, if it ever was.[7]

Whether Glasse took McConnell's research seriously isn't known, but he and Lindenbaum eventually did take seriously the notion of the kuru-cannibalism connection, and somewhat after the fact, in March 1967, he presented what he thought was his best evidence to the New York Academy of Sciences.[8] After a brief backgrounder of the disease, he laid out three main principles:

1. His bottom line was that women and youngsters of both sexes routinely ate the bodies of kuru victims, and they were the ones who got the disease. He elaborated a little on this first point: among the South Fore, many more women than men practiced cannibalism. The Fore said this had been true from the beginning of their involvement in cannibalism, somewhere around the start of the twentieth century. Although some experts claimed that the men participated but restricted themselves to flesh, not brain, Glasse argued that men avoided the feasts at least partly because they believed that eating human flesh made one susceptible to enemy arrows. (Glasse said he couldn't find a single man who would admit to having been a cannibal after the age of eight.) By contrast, in other groups where kuru was rare, males were cannibalistic much more often, although even they wouldn't eat the body of a woman—virtually no adult male would, on the grounds that women were somehow dangerous.

2. The way the body was cooked didn't help: putting human flesh potentially full of infectious agents into a

bamboo tube (with salt, ginger and vegetables) and then
steaming it at the altitude of the New Guinea Highlands
(where water boils at only 95 degrees Celsius) is, you
could say, a recipe for disaster.

3. When cannibalism stopped, almost immediately the
number of young children with kuru plummeted.

Nobody really questions that Glasse was the man who put
kuru and cannibalism together and backed up the claim with
observations. Others had speculated, but idly—while Carleton
Gajdusek, in effect, knew too much about kuru and so was re-
strained from leaping to the same conclusion for that very reason. Some (Gajdusek among them) apparently even had their
noses out of joint that it wasn't they who came up with the
answer.

In a way, Glasse's realization was typical of the way people
find solutions to perplexing problems. The history of science,
and it's particularly true in this case, gets boiled down over
time to the simplest and most straightforward version of how
things happened. But unlike an executive summary of a report,
which can fairly represent the contents, these histories present
a distorted version of how science works. It is never a proces-
sion of events, every experiment another brick in the wall. It's
more like this: put together an anthropologist thinking about a
medical mystery, a half-page article in *Time* magazine and an
idea that brings laughter when it's first suggested, follow it up
with serious thought, and you've got new science.

There are always ironies. Gajdusek knew more about kuru

than Glasse did, but Glasse read about James McConnell and made a weird and half-joking connection. In the end, Glasse was right. It might have been the only positive impact James McConnell had on science.

So cannibalism really did play a crucial role in spreading the disease, but where did the disease come from in the first place? According to the Fore, it was not even half a century old. Cannibalism could spread it but not start it. There had to be an index case, a "patient zero," for kuru. Even today that question has not been unambiguously answered, but the best guess is that a single Fore individual might have come down with what we now know to be a related affliction, Creutzfeldt-Jakob disease. It strikes, on average, one in a million people. It's mostly sporadic, in that there is no apparent cause, although some individuals are genetically predisposed. One in a million is not very common, and the susceptible New Guinea population was very small, but it's still possible that there could have been that perfect coincidence of a rare disease striking a single person at almost exactly the same time cannibalism began. If that person was eaten when he or she died, that could have started the epidemic. And as we will see later, Creutzfeldt-Jakob disease has played an important role in the evolving story of kuru. And vice versa.

3

Cannibalism
An Answer Guaranteed to Stir Things Up

Could there be a more spectacular and intriguing debut for what would become a new science? It had everything: an (anti) hero, a puzzling and fatal disease, an inspired guess and, above all, cannibalism.

The antihero, Carleton Gajdusek, never seemed comfortable with the idea that cannibalism had fueled the kuru epidemic. First, he dismissed the idea that Robert Glasse's linking of kuru with cannibalism was really any sort of achievement: "We are often asked when and how we first came upon the idea that cannibalism was involved in the spread of the disease. It is useless to speculate about the origin of this idea; I know of few Europeans who did not arrive at such a conjecture. All the missionaries, traders, miners, and government workers and their families in the Eastern Highlands knew that most of the indigenous peoples had been cannibals. . . . Europeans thus often suggested that 'cannibalism probably spread the disease.'"[1]

Gajdusek, with the exception of his obscure suggestion that infants might, through cannibalism, become sensitized to human brain tissue and then react to later consumption of it, continued to emphasize what he called the "complete ob-viousness" in an apparent effort to devalue Glasse's contribu-tion. It is ironic that Gajdusek, a medical expert who, seeing no signs of infection, could not or would not take cannibalism seriously, would nonetheless deride the *idea* of cannibalism as "obvious."

Funnily enough, even years later, Gajdusek persisted in seeking alternatives to cannibalism, for which, he insisted, there was no definitive evidence. He took the idea of acciden-tal exposure—bits of brain tissue being smeared on hands and faces and then gaining entrance to the body—and refined it to a scenario of mourners *deliberately* exposing themselves by a set of actions: rubbing their eyes, scratching and even cutting themselves with hands and knives soaked in brain, all as part of ceremony. Yet, the anthropologists on the scene, like Robert Glasse and his wife, saw no evidence of such ritual smearing at all, nor did the Fore talk about it.

Of course, adopting cannibalism as an explanation for any-thing is provocative: the practice is sensational, controversial and unsettling. So unsettling, in fact, that there are those who don't believe it ever had anything to do with kuru. They do have one piece of evidence or, rather, one piece of *missing* evidence, on their side: even though almost every account you read of kuru highlights the relationship between the disease-causing agent and eating human flesh, the fact is that no Westerner, no anthropologist, medical scientist or even mining prospec-tor actually saw the Fore people eating one of their own at a mortuary feast. It was all hearsay or testimony gathered in in-terviews. This missing piece has created a small but intensely argued academic cottage industry that asks whether it is really true that the Fore ate each other at all. Taken to the extreme, this line of questioning raises doubts about whether cannibalism itself exists now or has ever existed.

Those anthropologists who dare to ask that question, who dispute the very existence of cannibalism, acknowledge that there are instances when people do eat people—but only if they are deranged killers or are starved to the point where they will commit such a desperate act. Instead, their target is the supposed habitual cannibal, a member of a tribe that wins a battle, then executes and eats the prisoners, a manhunt in the most literal sense. They even distrust the claims of low-key secretive ritual cannibalism like that of the Fore. The most prominent of these skeptics, anthropologist William Arens, argues there is no real evidence that this has taken place. Ever. It's all just the same old sensational secondhand tales told by one culture (usually, but not always, Western) about another that they consider to be primitive. It's always the *others* who are cannibals. Arens does not claim that cannibalism has never been practised, nor that it never *could* happen. But he does say that his search of all the well-attested accounts of cannibals—from the Aztecs to the Hawaiians (who are supposed to have consumed Captain Cook) to Stanley's ("Dr. Livingstone, I presume") Africans to the Fore—fails to uncover any convincing eyewitness accounts, the kind he says you'd expect, or even demand, from anthropologists.

So, for instance, Arens points out that the detailed descriptions of the Fore's funerary practices ("the muscle stripped from arms and legs, the bitter-tasting gall bladder carefully cut away from the liver") were never actually witnessed by any Western observer.[2] By the time the first anthropologists arrived in the Eastern Highlands, the practice of cannibalism had been banned by the Australian administration, and it was gradually being abandoned. And if it hadn't been, the cannibalism that continued was wisely kept from prying eyes. The notion of cannibalism was entirely secondhand. In his book *The Man-Eating Myth,* Arens even argues that Gajdusek, in a reprint of his Nobel Prize acceptance lecture in the journal *Science,* misled readers by presenting two photographs, the first of a Fore woman dying of kuru, the second of a feast. One would presume that the two were directly connected, but the problem is, says Arens, the people in the second photo are feasting on a pig, not on the woman.

The revulsion most of us feel at the idea of eating human flesh makes the accusation that others are cannibals a particularly powerful—and common—form of mudslinging, and Arens thinks that too many anthropologists, who should know better, have fallen for it. He is undoubtedly right that many of the reports back from the Caribbean by Columbus's shipmates and from Africa in the nineteenth century claiming attacks by cannibals were based on rumor, prejudice or embellishment, or a combination of all three. It's like the Italians calling syphilis the French disease and the French returning the favor by calling it the Italian disease. But there's something about cannibalism that turns reasoned academic discourse into something much more heated. Take, for example, the case of the Aztecs, the powerful civilization that was ruling the area around Mexico City when the Spanish arrived, led by Hernando Cortés.

Spanish record keepers described ghastly scenes, including one where captured Spanish soldiers were taken to the top of step pyramids, laid on their backs and had their hearts cut out while still alive. Such sacrifices were not uncommon, at least according to the evidence gathered by the Spanish both from their own eyewitness accounts and from the testimony of Aztec nobles. Apparently, after the heart was removed, the corpse was pushed down the steps of the pyramid to the bottom, where it was butchered. The head took its place in a rack of skulls, the limbs were eaten in a stew by family and friends of the soldier who had captured the unfortunate prisoner and the torso supposedly went to the royal zoo to feed the animals.

If the only account of this was the Spaniards' own there would be ample reason to be skeptical, but there have been archaeological discoveries of headless human rib cages and piles of human skulls, which would attest at least to the fact that corpses were dismembered. But it was a modern analysis of this practice that really heated up the story. In the late 1970s, an anthropologist named Michael Harner did an accounting of the practice of Aztec cannibalism and came up with some staggering numbers: twenty thousand individuals a year were sacrificed across the Aztec empire, far more than any other known civilization. And the figure could have been higher: some estimates put the number of sacrifices at a single event, the dedication of the main pyramid at the capital Tenochtitlan in 1487, at around eighty thousand. Harner argued that there must have been some powerful driving force for cannibalism beyond the usual celebration of military victory. He theorized that the Aztecs needed to eat humans to make up for a protein deficiency in their diet.

His explanation was a biochemical one: as the population grew in Meso-America, any large animals that could have been domesticated or hunted were wiped out, and even though the Aztecs were skilled farmers, their lack of animal protein made them very vulnerable. The Aztecs could have acquired the eight essential amino acids, the building blocks of proteins, from corn and beans, provided there were no failures of those crops, and provided they were eaten simultaneously. When either of those crops did fail, they had nowhere to turn, said Harner, but to human meat.[3]

Just to put a final gruesome touch on the argument, Harner suggested that cannibalism was one reason that, unlike warring Europeans, the Aztecs didn't incorporate defeated enemy tribes into their empire. They preferred to continue to attack them and sacrifice them. It wouldn't have been acceptable to do so if they had been "Aztecs" themselves. So, in Harner's memorable phrase, those outlying areas were in a sense "stockyards."

Harner's theory didn't slide under the radar. Anthropologists who were already skeptical of the idea of institutional cannibalism were stirred to new heights of indignation when confronted with a theory that explained eating human flesh in terms of ecological needs. For many, Harner had gone too far, had suggested something too radical, and his claims were fiercely opposed.

The opposing views—cannibalism is real versus cannibalism is propaganda—are well entrenched. So it really is no surprise that when cannibalism was identified as the vehicle for spreading kuru among the Fore, there would be dissent. William Arens, having claimed there was no hard evidence that the Fore practiced cannibalism at all, went further: while it was not impossible that some tribes somewhere, sometime, were cannibals, he had yet to see the evidence.

But isn't the weight of the evidence, however circumstantial, on the side of cannibalism among the Fore anyway? Certainly, those Westerners who first encountered the Fore people in the mid-twentieth century, the miners and adventurers swapping stories in the bar about cannibals, shouldn't be taken as witnesses for the Crown, but what about the anthropologists and medical personnel? Were they, too, simply taken in by the myth of cannibalism?

If they were, they sure weren't aware of it. Anthropologist Annette Beasley conducted a set of interviews with the Fore years later, and her translator related this version of one old man's comments: "X [is] fearful about matters spoken with respect to the connection between kuru and cannibalism—[the] younger generation do not like to hear it talked about and would chastise him."[4]

Vincent Zigas, one of the first medical experts to chase down kuru, wrote: "Not long ago in a remote hamlet I had encountered a group of adolescent males masticating small pieces of dark, smelly, fleshy substance. My enquiry revealed that they were chewing the flesh of a recently deceased great warrior. I requested to see the body, and found it placed on a sugar-cane-covered bamboo platform. Decay had already set in. The dissection of the pectoral, deltoid, gluteal, bicep and quadriceps muscles was performed with superb skill."[5]

Zigas told them he was hurt that the Fore hadn't told him about ritual cannibalism. In response, they pointed out that he hadn't asked. Carleton Gajdusek made his own references to cannibalism having been common before his arrival, and it is said that he had a video, although none has ever been publicly shown. To disbelieve accounts like this, you have to distrust either the writers or their sources. But in the singular case of cannibalism, that commonly happens.

Nonetheless, the anthropologists on the scene, as well as the medical experts like Gajdusek, harbored absolutely no doubt that the Fore had been practicing cannibalism since the beginning of the twentieth century. The issue wasn't whether or not they were doing it, but what role it might play in the spread of kuru. In the end, after all the alternative hypotheses had been found wanting, cannibalism took its place as an essential piece of the kuru puzzle.

But where did that leave the cannibalism skeptics? Unwilling as they were—and are—to give ground and accept mere testimony with no accompanying eyewitness accounts, they have resorted to finding loopholes in the cannibalism argument.

One of those loopholes was suggested by Carleton Gajdusek himself. From time to time he went on record as saying that it wouldn't have been necessary actually to eat kuru-infected tissue. As I mentioned above, he seemed to favor the idea of self-infection by imagining the Fore deliberately smearing themselves with infected brain tissue. There were also plenty of opportunities during the preparation of bodies to become accidentally infected with the causative agents by contamination of hands first, then the eminently infection-susceptible mucosal membranes in the eyes, nose and mouth. There was no hand washing, no attempt to prevent such potential infection. The chance for infection would probably last several days.

But there's really no way of knowing if haphazard exposure like this could account for the numbers of deaths. It's likelier that actually chewing and swallowing infective material would be more efficient.

But there have been other attempts to exonerate cannibalism. Lyle Steadman and Charles Merbs, writing in *American Anthropologist,* argued that the handling of skulls of the recently deceased, not cannibalism, was the route of infection for kuru.[6] The advantage of this notion, according to them, was that at least the handling of skulls, unlike cannibalism, was more than a rumor. Westerners had actually seen it happen.

Here's what they proposed: some New Guineans living in areas around the Fore were known to keep the body of a deceased relative around for days after death, in some cases allowing the putrid flesh to drip on the widow's cape, thus signifying her as being in mourning. Others buried the body, then, a month or two later, retrieved the skull and hung it on a wall inside the house. Steadman and Merbs showed a picture of a woman threading a piece of cane through the eyeholes of a skull dripping with decomposing brain and lifting it out of the ground. There is no doubt that handling a diseased skull and brain could in this way have transmitted kuru. The women who handled the skulls would then, of course, have to touch the young children with their infected hands in order to spread the disease to them. There is, however, one significant problem with their argument. Although there is convincing evidence that some groups in New Guinea did this, there is not, according to Steadman and Merbs, any direct evidence that the Fore handled their dead in this way—only the testimony of anthropologists who wrote that often bodies were buried before being exhumed and consumed.

But wait a minute: if we're supposed to be skeptical of anthropologists' claims of cannibalism, why should we believe other anthropologists' stories of burying and unburying? Either they are fair witnesses or they're not—it's not justifiable to cherry-pick from testimony. So while this skull-handling idea could have some merit, as far as evidence goes, it's no better than cannibalism.

Finally, Steadman and Merbs, in an attempt to explain how kuru arose in the first place, take issue with the idea that a solitary case of the related Creutzfeldt-Jakob disease (CJD), coupled with cannibalism, might have triggered the kuru epidemic. Instead, they suggest that CJD might not have been present at all in New Guinea until introduced by a Westerner— a miner, missionary or patrol officer—around 1920. But this makes no sense at all: CJD is not communicable. Even if someone arrived in New Guinea and then died of CJD, there is no way they could have infected any of the Fore. Unless . . . heh, heh . . . the Fore *ate* that person. But that just didn't happen.

So launching the proper study of kuru required leaping over some obstacles: a reluctance to believe in cannibalism and an inability to accept that it might be the mode of transmission. And even when cannibalism won over the majority of researchers, kuru still remained an isolated curiosity, a strange disease, contagious by virtue of strange practices, in a remote corner of the world. There were no hints that it would soon have much wider—even global—implications. As brilliant as Carleton Gajdusek had been in bringing kuru to the attention of the medical world, new points of view were needed. It wasn't long before they appeared.

4

Igor and Bill

The Discoveries That Bring
Kuru to World Attention

Don't read a science textbook if you want to get a sense of how science really works. Real flesh-and-blood people aren't part of the story of science in textbooks; almost nothing is said about creativity, no unraveling of the stereotypical image of the scientist as clinical, detached, scholarly and not too social. It's no wonder kids turn off science in high school.

While there are famous instances of brilliant scientists working entirely on their own, that is not the norm. Teamwork is, and unless there is a critical mass of minds focused on a problem, it can go unsolved.

In the beginning, there were many reasons kuru fascinated those who investigated it. It appeared to be a unique disease. It was exotic, confined as it was to one particular group of what Westerners liked to call "stone-age" people in New Guinea. It was certainly mysterious: a disease whose cause could not be identified and whose mode of transmission couldn't be found. And not least, it aroused compassion in all who were there in New Guinea to witness it: young children as well as adults were succumbing to an ailment that robbed them of their ability to walk, then stand, then even eat, while at the same time allowing them to be aware of all that was happening. And it was invariably fatal—in the few reported "remissions" of kuru, it turned out that the patient didn't have the disease in the first place.

Kuru had no broad medical application as far as anyone knew. It was certainly fascinating (Sir John Eccles, an Australian Nobel laureate, thought it was the most interesting medical puzzle he had ever come across), but New Guinea was far from the centers of science and medicine, and, once you were there, to study kuru was arduous at best. The disease was unyielding, the conditions lousy. It was uphill for Carleton Gajdusek and anyone who worked with him. It merited international attention but wasn't accessible to it. Kuru needed more eyes on it and more lab technology applied to it. Somehow it had to leave the jungle and reach the labs, and that is exactly what happened. Two discoveries brought kuru to the rest of the world, and both were due to the keen eyes and agile minds of people who had never seen a kuru patient in person nor had even been to New Guinea: Igor Klatzo and Bill Hadlow.

It took Gajdusek some time before he was able to persuade the Fore to allow him to perform autopsies on those who had just died of kuru. Once he had won their confidence and their assent, he sliced the brains using the crude tools that were available to him, preserved those slices in formaldehyde, froze them and shipped them back to Washington. There his colleague Igor Klatzo, an expert in the fine structure of the brain, did the final preparation and scanned the brains under the microscope.

Klatzo had no trouble finding abnormalities in kuru brains—

making *sense* of those changes was the issue. It's important to appreciate that while the brain is often likened to a computer, the internal layout is totally different. Computers are geometric, vast micronetworks of lines, squares and rectangles, laid out with precise forethought. Brains evolved with no forethought at all, just the imperative of survival. They are built on the past, the end product of millions of years of evolutionary tinkering with circuits of neurons. Unfortunately, digging through the brain is not archaeology—there are no precise strata that can be dated to a particular epoch. People love to refer to the "reptilian brain," but there is no such easily identifiable set of structures or behaviors associated with reptiles in the modern human brain (nor did Paul MacLean, the man who coined the term, intend it that way). Neither does it make sense to attribute all kinds of human abilities to one or the other of the two cerebral hemispheres; right and left communicate with each other at lightning speed every second of the day.

The brain at the microscopic level is a dense interwoven set of networks involving many different kinds of cells: there are perhaps one hundred billion neurons, the actual signaling devices, but they are accompanied by many others, including an even vaster number of glial cells, whose role is just now beginning to be understood. There might be ten glia for every neuron, and the list of what they're responsible for grows yearly.

When Igor Klatzo received brain tissue from kuru victims, he couldn't just put a slice under the microscope and check it out: the picture would be too bewildering. He had to use a variety of stains to highlight each kind of cell to get a sense of what he had before him.

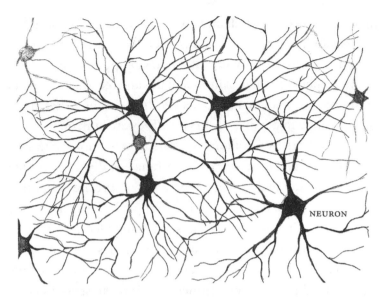

NEURON

Take the thinnest slice of brain tissue you can, put it under the microscope, magnify it several thousand times, and individual neurons can
be resolved, each with multiple extensions reaching out to other neurons. But these are just the neurons and, at that, only a few of them:
there are many more neurons, and countless other kinds of cells that
have been left out for clarity.

What he saw was chaos. Neurons had degenerated and
taken on weird and distorted shapes; the star-shaped astroglial
cells had proliferated wildly, and there were odd, dark deposits, similar to the so-called senile plaques found in the brains
of those who died of Alzheimer's disease. In some cases, even
the brain's blood vessels were dilated and small hemorrhages
had occurred. There was more damage in the cerebellum than
anywhere else. That made sense: the cerebellum is primarily
a motor control and coordination center, and the hallmark of

kuru was a steady decline in the coordination of movement. The cerebral cortex was largely, though not entirely, spared, and this too was consistent with the fact that kuru victims were intellectually intact until near death.

Klatzo's first full report in the journal *Laboratory Investigation* was a full forty-eight pages—some of the individual case reports took up twelve pages of pictures and text. But in all that detail, there were apparently no revelations. "I am afraid I am unable to give you any useful leads as far as the etiology of this disease is concerned," Klatzo wrote to Gajdusek on September 13, 1957. "It seems to be a definitely new condition without anything similar described in the literature. The closest condition I can think of is that described by Jakob and Creutzfeldt. However, in cases reported, the neuronal degeneration was most intense in the cerebral cortex; also, no cases involving children or adolescents are mentioned. The etiology of Jakob-Creutzfeldt condition is entirely unknown (only about 20 cases have been reported). Since there are no known hereditary conditions resembling even remotely the kuru condition, I am inclined to suspect some toxic metabolism for the process."[1]

Klatzo didn't realize it at the time, but rather than failing to identify any useful leads in identifying the etiology, or cause, of kuru, he had actually started to crack the mystery. That similarity to the "Jakob-Creutzfeldt condition" he referred to established the first link between kuru and another human disease. He was being appropriately cautious, noting that, in that disease, the cerebral cortex—the thinking brain—was hit hardest, while being largely spared in kuru. Also, the youngest

patient recorded to that time was twenty years old, whereas
kuru victims were often younger than ten. And really, caution
was appropriate when you realize that he was comparing kuru
to a disease that, first, was not well known at all and, second,
exhibited huge variation even among the cases that had been
identified.

The disease Klatzo referred to is now generally called
Creutzfeldt-Jakob disease, or CJD. Here's a pocket history: in
1921, Alfons Maria Jakob described a previously unrecognized
neurological deterioration in five patients, male and female,
ranging in age from their mid-thirties to mid-fifties. It was a
devastating condition that robbed them of their mental life
and memory, eventually leaving them unable to walk, talk or
speak. They died, completely demented, within a year after
the appearance of the first symptoms. In one of the reports he
published on this disease, he concluded that another patient
described in 1920 by Hans Gerhard Creutzfeldt had had the
same disease, and so he felt compelled to acknowledge the earlier description, coining the term "Creutzfeldt-Jakob disease."
It turned out he was being unnecessarily generous, because recent analysis of the microscope slides of all the patients reveals
that although Creutzfeldt's patient definitely had something,
it wasn't the same disease as that affecting Jakob's group. In
fact, three of Jakob's five didn't have the same disease either! As
a result, there's been the suggestion that the name should be
switched to Jakob-Creutzfeldt disease, but it doesn't look like
that's going to happen.

Establishing the link to CJD was Igor Klatzo's major contribution to unraveling the kuru mystery, but he did something

else as well. His micrographs of kuru-damaged brains made their way into the reports that he, Gajdusek and Vincent Zigas published in the scientific journals. Some of these same photos were also put on public display in a traveling exhibit on kuru that eventually made its way to England. And it was there that a second, hugely important breakthrough was made.

In 1959, Bill Hadlow was in his late thirties and, at least temporarily, had come to work in England from the United States. Hadlow was a veterinarian with a special interest in the signs of disease in brains, and he was assigned to a lab in Compton that specialized in a sheep disease called scrapie.

Unlike kuru, scrapie was not a new or unknown disease—the first definitive records in England go back to the transcripts from Parliament in 1755.[2] Wool was a big money-maker at the time: there were ten million sheep in England, and a quarter of the human population was employed in the production of meat and wool from those sheep. The Industrial Revolution was near, and technology like the flying shuttle was already beginning to move wool-working from the cottage to the factory.[3] Farmers and entrepreneurs alike were understandably concerned with the spread of a mysterious disease among flocks of sheep in the English countryside. The issue made it to Parliament at least partly because farmers were suspicious that the jobbers, the men who ran the sheep trade at the time, were trying to get rid of their diseased sheep by mixing them with healthy ones, hoping they wouldn't be noticed.

The disease was something that sheep farmers had never seen before. For no apparent reason, sheep in flocks here and there would suddenly become skittish and clumsy, sometimes prancing, sometimes walking with their front legs but trying to gallop with their hind legs. They had a bewildered look, and much of the time appeared to be suffering from some sort of intolerable itch, rubbing against tree trunks or fences until they were virtually naked. The disease was invariably fatal, but the cause, especially in those early days of biology and medicine, was unknown. (As late as the mid-nineteenth century, lightning and sexual excess were still being proposed as triggers.) England wasn't the only country where sheep were coming down with this weird affliction: a report from about the same time in Germany noted that scrapie appeared to be contagious, necessitating the isolation of affected animals as soon as symptoms were noticed. (The fact that such animals were apparently infected with something deadly didn't appear to faze German landowners, if the advice in the same docu-ment is any indication. Those sheep should be slaughtered, the document says, so that they could be consumed by the noble-men's servants.)

The idea that scrapie was an infectious disease, although noted in the early German writings, was really a minority opin-ion through the eighteenth and early nineteenth centuries: most thought it was an inherited condition and that somehow it had arrived, at least as far as the English were concerned, with the first imports of the fine-wooled merino sheep from Spain. For reasons that aren't really clear, opinion began to swing back toward infection by the middle of the nineteenth century, but not because anybody had demonstrated that it was infectious, by, say, transmitting the disease from sheep with scrapie to those without. Around that time, French scientists tried to do exactly that: inoculating sheep with extracts from the brains of animals that died of scrapie, and housing affected sheep with healthy ones, to demonstrate the disease's transmissibility. But even though they waited months for symptoms to show up, none did. As we now know, they were *exactly* on the right track, but they had missed one crucial point: time.

It wasn't until the 1930s that another generation of French researchers showed that inoculating did work—it was just that you had to wait a year or two before symptoms appeared. *A year or two.* Compare that to a flu epidemic, where, if you're going to get it, you'll get it within weeks of the virus's arrival in your community. When flu outbreaks are ending after a few months, scrapie is scarcely getting started. Nonetheless, there was a strong consensus developing that, as slow as it was, some sort of virus had to be responsible.

By the time Bill Hadlow came to England to work on scrapie, the disease had spread to many countries, again via imported sheep. The disease was hitting sheep farmers hard, and there was a significant research effort ongoing, especially in France and England, to figure out what was causing it. But it was turning out to be a difficult puzzle to solve. The uncertain seesawing theories of infection versus heredity had not stopped, even though there had been some dramatic demonstrations of transmission of the disease, including an unfortunate case in which a vaccine against a different sheep disease was made using material from scrapie-infected sheep, with the result that the vaccinated animals were indeed protected against the disease targeted by the vaccine but died of scrapie instead. There had even been some pioneering attempts to characterize the infectious agent, and it might have been these studies that turned some away from the idea of infection, because the experiments implied that whatever was causing scrapie was totally weird: smaller than bacteria, resistant to all kinds of chemicals and extreme heat, and hardy as hell, capable of surviving and remaining infective in dried brain tissue for two years. There really wasn't anything that should be able to do that, but there it was.

Added to that uncertainty was the brilliant demonstration by English researchers, using an experimental design that should go down in history as one of the most ambitious ever, that different breeds of sheep were differentially susceptible to scrapie, ranging from 78 percent in Herdwick to 0 percent in Dorset Downs. The experiment was conducted by William Gordon, the same man who inadvertently infected sheep in the infamous vaccination campaign, and its claim for some sort of world-record status rests on the fact that more than a thousand sheep were involved, and in each sheep scrapie would take at least a year to appear—an experimental design on a scale of difficulty unimaginable by most scientists today who can work cell cultures or easy-to-maintain organisms like roundworms or fruit flies.

If you were looking for a time of excitement and change in a research area little known outside the walls of the lab, this was it. And Bill Hadlow was there. Applying his deep knowledge of neuropathology, he noticed at once that the brains of sheep that had died of scrapie exhibited not just the well-known holes in brain cells themselves but also many other signs of damage, including a spongy, perforated appearance of the brain tissue

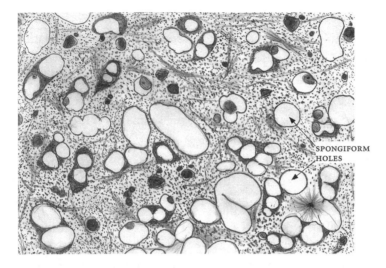

SPONGIFORM
HOLES

This slice of brain would normally be packed densely with tissue. But multitudes of prions have damaged the brain so much that holes are scattered everywhere, empty spaces all over the cellular landscape. That's why prion-caused illnesses are called "spongiform" diseases.

and an explosive proliferation of star-shaped cells called astrocytes. As Hadlow himself remembers, "I did not know of another disease like it, in man or animal."[4]

So the understanding of scrapie was progressing as you'd expect: slowly, unsteadily, hampered by the fact that what knowledge there was of a possible infectious agent was scant and mysterious. Then a seemingly random and inauspicious event intervened and the picture changed forever.

It was June 28, 1959. A colleague of Hadlow's, William Jellison, came for a visit from the United States. The previous day, Jellison had been in London, where he had gone to a medical exhibit at the Wellcome Historical Medical Museum. It was an exhibit about kuru, organized by Carleton Gajdusek. Jellison told Hadlow he might find it interesting, being a brain pathology guy and all, and a few days later Hadlow took the train to London and checked it out. For Hadlow, the exhibit was more than just interesting—it was astonishing. There, in front of him, were images of the diseased brains from kuru victims, and the brain cells had the same holes—the vacuoles—that he had seen so often in scrapie brains. While in London, he gathered all the articles he could on kuru, then returned to his lab and for the next few days thought about little else but the resemblance between kuru and scrapie. It wasn't just the holes in neurons; the two diseases were similar in nearly as many ways as two diseases (one in sheep and one in humans) could be. He described the resemblances as "uncanny," and wrote a letter to the medical journal *The Lancet* listing them.

Even though Hadlow began the letter by cautioning that he was in no way suggesting that the diseases were "identical or even counterparts," his list of resemblances is so long that the more you read, the more you wonder why it had taken so long for anyone to stumble on this. Each disease was confined to certain populations, whether tribe or flock. The victims could come down with the disease even months after being removed from those populations. The cause of both was obscure, but genetics could play a role, whereas nutritional or toxic factors apparently didn't. Both diseases were inevitably fatal after several months of deterioration, especially of movement control and coordination. And, of course, there were the hallmarks of disease in the brain itself, the symptoms that captured Hadlow's attention in the first place. After laying out these parallels at almost every level of both diseases, Hadlow made the crucial point. Even though no one knew exactly what caused scrapie, the fact was that it could be transmitted between animals by inoculating with ground-up brain matter: "Thus it might be profitable, in view of veterinary experience with scrapie, to examine the possibility of the experimental induction of kuru in a laboratory primate."[5]

Hadlow's letter was transformational, although the route taken by that transformation is a little hazy. Worried that a printer's strike might delay its publication in *The Lancet*, Hadlow sent Carleton Gajdusek a copy in the mail. What then followed is confusing. Gajdusek does not record having received Hadlow's letter (though he was a prolific recorder of everything he was thinking and doing), but he did reply to it. That reply wasn't published until 1992, however, and Hadlow does not remember receiving it back in 1959. In his reply, Gajdusek asserted that he and his colleagues were pursuing the idea of infection but that "we have thus far had poor luck with inoculation experiments . . . frozen and fresh materials are being injected into a number of animal hosts during this year's work on kuru."[6] This is pretty strange—what animals? In what lab? There is no evidence that such experiments had started, not even any evidence that Gajdusek had even *thought* of doing such experiments. Nevertheless, Gajdusek and his colleagues did begin inoculating chimpanzees with extracts of kuru-infected brains, but not until August 1963, four years after Hadlow suggested it. And as Hadlow anticipated, they had to be patient: the first symptoms of a kuru-like disease didn't appear until June 1965. But they did appear. And a couple of years later, similar experiments showed that Creutzfeldt-Jakob disease, the condition that Igor Klatzo had recognized as being similar to kuru, could also be transmitted to primates. Everything had changed.

Hadlow's letter, less than a single page in length, was the agent of that change. But in remembering the sequence of events, he downplayed the importance of his role: "Over the years, much has been made of my letter pointing out the likeness of the two diseases and its implications. That, of course, pleases me. Yet as the events bear out, the observation was largely fortuitous. Certainly knowing something about scrapie, especially its pathology, helped me to see the resemblance at first glance. Nevertheless the likeness of the two diseases is such that no doubt sooner or later someone would have become aware of it."[7]

Hadlow is no doubt right that someone else would eventually have linked kuru and scrapie. But does that somehow lessen the importance of his contribution? You get the feeling from his comment "sooner or later someone would have become aware of it" that he would reject any suggestion that his personal style played a role in boosting the influence of his journal letter linking kuru and scrapie, but it stands as a near-perfect example of the combination of scientific writing and thinking. It was short, sharp and to the point, it listed the similarities between scrapie and kuru succinctly but completely, and it proposed the crucial experiments of inoculating animals with material from kuru brains. All this not from an expert in human disease but from a veterinarian! That a vet could make this kind of contribution is what apparently pleased Hadlow the most.

You might have thought that once the inoculation experiments to test the infectivity of both kuru and CJD were under way, and en route to success, the mystery of these diseases, together with their unlikely cousin, scrapie, would be short-lived. Unfortunately, nothing could have been further from the truth. Sometimes clarity brings complexity, and that was certainly true in the 1960s with these diseases. In a way, the research took off in two different directions. Once it had been shown that you could infect animals with CJD, and that after a period of months they came down with a version of the disease startlingly similar to that in humans, medical researchers started to ask questions about the epidemiology of the disease. One in a million people worldwide get Creutzfeldt-Jakob, but what does "get" mean? Is it truly sporadic, with no apparent cause? Are some people genetically prone to CJD? Or, because it had now been demonstrated to be infectious—at least in the lab—were those so-called sporadic cases actually the result of infections? Those questions drove the human thread of the research.

But there were also questions about the nature of the infectious agent. And because scrapie was much better known and already had a lengthy record of investigation, the animal thread of this story became the search for the scrapie agent. To appreciate why, we have to shift our gaze from the macro to the micro, down into the cell itself, to see exactly how it can be attacked by infectious agents, including bacteria, viruses and now this most insidious agent, the cause of kuru, scrapie and CJD.

5

The Life of a Cell
A Miraculous, and Often
Precarious, Complexity

We've already encountered cells like the neurons and glia in the brain. It is their disruption and death that is the central mechanism underlying the diseases we are pursuing: kuru, scrapie and CJD. But to appreciate fully just how dramatically different from all other maladies these diseases are, and to set the stage for the revolutionary science that was soon to emerge from the early period of exploration in the 1950s and 1960s, we have to appreciate the complexity of these cells. Not so much as a biochemist would do it but as a biotourist might, gaining at least a mental snapshot of what a cell is really like.

We'll start in England in the 1660s, an amazing time for science. Isaac Newton had returned to the family farm to avoid the plague at Cambridge, and as he lolled around in the backyard he happened to see an apple fall from a tree and somehow, amazingly, linked the descent of the apple to the orbit of the moon around the earth. Newton wasn't in London much, but he belonged to the Royal Society, a club of übersmart scientists which began there in 1660. It was an incredibly select group, and while it still exists in name, it would be hard to argue

that the current version is equal to the original. Besides Newton, Christopher Wren, the architect, surveyor and creator of
St. Paul's Cathedral, was a founding member and later president (years later he would be replaced in that role by Samuel
Pepys, and, many years after that, by the aging Newton himself); so were the great chemist Robert Boyle and John Aubrey,
the man who described and mapped the megalithic monu-
ment at Avebury. But my favorite member of the Royal Society
was Robert Hooke, the underrated, overshadowed, multital-
ented, hard-nosed guy who dared to get into a serious feud
with Newton.[1]

It's likely you've never heard of Robert Hooke. Even though
there have been several books about him, they've made little
impact: Hooke will never get the credit he deserves. Rather
than reviewing a long list of his accomplishments, just note
one thing: he was appointed by the Royal Society to be curator of experiments. The curator was obliged to perform three
or four "considerable" experiments for the pleasure of the assembled scientific and technological geniuses every time the
society met, which was usually *every week*. Remarks in the society's proceedings to the effect that Hooke was "ordered" to
bring in two or three good experiments for the next meeting
were not uncommon.

One of the many significant discoveries Hooke made—this
one with his own handmade microscope—was the tiny pores
in a thin slice of cork. He called them "cells," and he was pretty
sure no one had ever seen them before. The cells that Hooke
sketched in his book *Micrographia* were empty rectangular shells,
the walls of cells long dead. Apparently, they reminded him of

the minimalist rooms in monasteries, the cells of monks. So although Hooke gets credit for understanding that these were fundamental units of living tissue, he had no idea of the complexity he had just begun to reveal. His technology certainly wasn't up to the task of such revelation, nor was the state of biology at the time.

But as much as our concept of cells has changed, the name has stuck. Unfortunately, the word "cell" suggests emptiness. That is very misleading. It contributes to a generally held idea that cells are, if not empty, simple, full of some sort of bland fluid with—at most—a few objects bobbing around in them.

They are anything but empty; nor are they simple. Indeed, they are so alien, so remote from everyday existence, from any form of life we see in the macro world, that words can't do them justice, and the most accurate pictures, if faithful to the crowdedness of the cell interior, are just too difficult to follow. In the words of L. L. Larison Cudmore: "Cells have everything. But visibility."[2]

Cells are what you make of them. If you're interested in architecture in miniature, you'll notice that the entire volume of a typical cell is filled with structures of elaborate design but brilliant functionality. If you lean toward engineering, you can't help but be entranced by the cell's high-volume industry and energy production. If you're into perception, or signaling or stimulus-response apparatus, just look at the confining membrane of the cell, with its array of sensors that allow it to monitor everything that is happening outside and respond if necessary.

This is all about survival and reproduction, the elements of evolution. Most cells run through the reproductive cycle (so cheerlessly memorized in so many biology classes), called mitosis.[3] I'm sure somewhere in your memory—if not actively repressed—are images of chubby chromosomes collecting at the middle of the cell, then separating and moving to opposite ends of the dividing cell. The DNA in those chromosomes is the software for building and running the cell. A "blueprint" is a common analogy, but it's not dynamic enough. Rapid prototyping, where computer code is used to construct objects layer by layer, is a little better, but none of these inanimate comparisons really works.

For instance, here's what's involved when one molecule is made in a human cell, just one of the thousands or even hundreds of thousands that are synthesized every moment. Let's say it's a protein. That means it, at least by molecular standards, is huge, a chain wrapped around itself several times, forming loops, spirals and sheets, the entire thing held relatively steady by electrical and chemical forces. Let's further say that once this protein is made, it is destined to move through the cell and plug into the outer cell membrane, a flexible oily layer that separates the interior from its surroundings, preserving the integrity of the cell by doing so. Once there, the protein will act as a receptor for signaling molecules arriving from other cells in distant tissues.

The story starts in the cell nucleus. First, the gene that codes for that protein has to be activated. That requires the action of a variety of molecules that find the beginning and ending of that gene somewhere in the vast library of DNA, mark it, then arrange for it to be transcribed into a short-lived copy made of its analogue RNA. That piece of RNA moves through a portal in the nuclear membrane like a slow-moving rocket exiting the giant mother ship.

Once outside the nucleus, the RNA finds and attaches to the elaborate machinery that will read the information stored in it and use that information to assemble the protein, bit by bit. The RNA might run through several sets of such machinery to create multiple copies of the protein.

That is the basic process. Of course, that description is so simple it's unreal, and it stops short of the really interesting stuff. For instance, that piece of RNA? It is a faithful copy of that length of DNA—of the gene. But the gene has pieces in it that are not part of the code for the protein, and these need to be edited, snipped out before the RNA is read. There can be several of these noncoding pieces, and the RNA has to visit a special object called a spliceosome to have that surgery done. What's more, the code is so adaptable that a single piece of RNA can have different sets of these no-coding segments removed, with the result that a single original RNA chain can yield several different versions (after trimming), which in turn will make several different proteins. That is really just amazing. Imagine being able to take a single English sentence, cut out a word here, another couple there and then a phrase here and still have it make sense in French. Then do it again, throwing out different pieces and have it make perfect sense yet again. That is what this amounts to.

If this were a treatise on the complexity of cells, we could go off in all directions right here and now. For instance, what is the point of these pieces of so-called junk DNA? What good does it do to have a perfectly good gene interrupted here and there by stretches that not only don't contribute at all to the protein that gene makes but also require part of that very gene to be devoted to their removal?

Could pieces of junk DNA, in fact, be so valuable that they define the difference between humans and chimps? One of the reasons we are so different from chimpanzees—despite famously sharing 98 percent of our genes—is that the ways chimp and human genes are spliced and pasted together again are very different, especially in those cases where there's an opportunity to splice out different pieces and thus create different readings of that gene and ultimately different proteins. Fascinating indeed, but back to our protein.

The DNA code specifies the sequence of subunits that make up the protein, but that's by no means the end of the story. The protein may need some decoration, additional clusters of sugars or other atoms, to make it complete. Then it has to get to its destination, and that could present the challenge of a lengthy journey through channels to the outer membrane. When it arrives, the protein moves into position, spanning that membrane so that its head is sticking out of the cell, its tail still safely inside.

Once there, it joins hundreds of others, similarly embedded in the membrane, monitoring an extracellular traffic jam, a bombardment of everything from fragments of atoms to giant molecules, some bearing crucial information, some a threat to the very life of the cell. If the right molecule happens along, the one that our protein has evolved to recognize, the two come together and their bonding triggers changes in the shape of

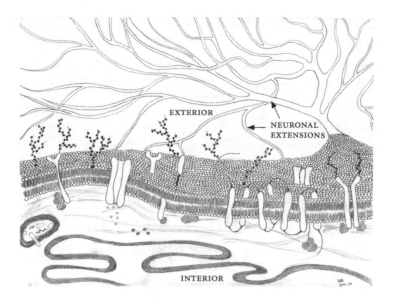

Magnify the outer membrane of a human cell a few million times and this is what you'd see: an elaborate array of sensing and signaling molecules adorning it. Imagine that this scene is actually dynamic, with molecules wobbling, shivering, flipping head over heels.

the protein that are in turn recognized by other molecules adjacent to it inside the cell. An informational cascade is kicked off, and the cell changes, maybe in a small way, as a result.

But the complexity of the cell is such that a description like this is not much better than Hooke's empty sketches. The cell has a skeleton that uses rods and ratchets that extend or contract to change the cell's shape or even allow it to move; materials are packaged and shipped from one place to the other using cables, like some ultramicroscopic zipline; and everywhere new molecules are being built as others are being dismantled.

Construction is only half the story. As long as oxygen is available, cells live by ripping apart sugar molecules for their energy. The final steps in this process take place in the sausage-shaped mitochondria. Mitochondria themselves are a wonder: billions of years ago they enjoyed life as free-living bacterial cells. Perhaps *enjoyed* is an exaggeration because when given the opportunity at some point in the very distant past to forgo their nomadic life and settle down as permanent residents inside a host cell, they took it, eventually adapting themselves to the intracellular environment by reducing their complement of genes to the point where, if separated from their cells, they cannot survive.

One more feature of the living cell: while the protein described earlier is a part of the cellular architecture, many more proteins act as facilitators, accelerating the pace of chemical reactions inside the cell. Without those proteins, known as enzymes, life on earth simply wouldn't exist.

Coming up with a mental image of all this is extremely difficult. There is no single correct way of envisioning the interior of a cell. The images that were prevalent in textbooks through the twentieth century, while true to the ability of the microscopes of the time, nonetheless gave the impression of a clean and orderly arrangement, but the true nature of things is very different. Orderly yes, but an order that wouldn't be easily discernible to us because the entire scene is so incredibly crowded. Time moves quickly—most of the cellular constituents have a very short lifetime. Manufacture is unceasing, movement continuous, life—even at this level—frantic.

In the end, intracellular life is best represented as a blur, since any effort to make that world perceptible is artificial. Do you color all the constituents? It helps, but they don't actually have any color. Do you represent each and every molecule as exactly as you can, with every single atom in place? That's problematic, because atoms are not clear-cut entities but blurred and shadowy; they are cloudy, not concrete. Even scientists prefer to represent molecules in broad strokes, making it possible to see the overall shape, locating the clefts and promontories, because it's this topography that determines what that molecule can and will do for the cell.

Life, even for something as finely tuned as a cell, can be precarious. While it is not a delicate thing (it defends itself, it corrects the inevitable errors that crop up when energy consumption and growth are running at capacity, it has redundancy), there are hazards. Genes can mutate, condemning the cell to any one of thousands of inefficiencies or other life-threatening shortcomings. Cancer can corrupt and kill. And even though it has history on its side—the cells in our bodies are the descendants of a lineage reaching back more than a billion years—so does the enemy. For that same billion-plus years, pathogens of many kinds have been seeking ways to circumvent cellular defenses, whether by subterfuge, piracy or out-and-out assault. When one side evolves a defense, the other evolves a new offense. After a billion years, it's not surprising that both sides operate at a high level of sophistication.

6

The Death of a Cell
By Subterfuge, Piracy or
Out-and-Out Assault

Until researchers began to try to identify what might be causing kuru, scrapie and CJD, viruses, among all the disease agents
that can incapacitate or kill a cell, were the most insidious and
intimate. On their own, viruses are nothing—not even living
things—but once they gain access to a living cell, they multiply
quickly and ruthlessly. The way they do this has been a constant
thread running through the story of these mysterious diseases.

Nobel laureate Sir Peter Medawar described a virus as "a
piece of bad news wrapped in a protein." Bad news it definitely
is, but as an example of how to subvert a living cell, it has no
peers. Here's the short story: a virus is nothing much more
than a package of genes, often only a handful, protected by
a suit of armor of interlocking protein molecules. Viruses are
extremely small, in the range of tens of nanometers (billionths
of a meter). There are elaborations on this structure, but that is
the basic model. If a virus particle is fortunate enough to gain
access to the body (and remember, it's inanimate and so only
can drift helplessly through the air or in fluids), it must come
into contact with its target tissue to do any damage.

Whether the cells of any particular tissue are susceptible depends on the array of molecules on their surfaces, the receptors that monitor the outside world and report back to the interiors of the cells. Viruses, as simple and unadorned as they are, relative to the cells they attack, display molecules on their surfaces that match certain counterparts on the cell surface, and allow them to gain entry through the cell membrane. This relationship is generally likened to a lock and key but in molecular detail is much more complex.

Once past the barrier of the cell membrane and inside the cell, the virus's protective protein coat is shed, its job done. The viral genes are inside and proceed to take over, shutting down the activity of the cell's own genes and turning themselves on, redirecting the energy and activities of the cell to one end: producing more viruses. Many more viruses. The cell becomes a viral factory floor, with key virus enzymes being made here, viral coats over there, copies of the viral genome here. A supply-chain expert would love it. Once all the pieces are made, they're assembled, one last viral protein is called in to breach the cell for the escape, and a new generation of viruses, sometimes numbering in the hundreds of thousands, escapes to infect neighboring cells. This is what going "viral" is really all about.

Just as saying that cars are four-wheeled vehicles with seats and engines says nothing about the difference between a Ferrari and a Smart car, the brief outline above describes a generic virus only. Each real one has its own variations on that theme.

So, for instance, the rhinoviruses, the group of two hundred or so that are responsible for more cases of the common cold than any other virus, aren't much more complicated than the description you've just read. They aren't just inanimate, they look it, being perfectly symmetrical twenty-two-sided icosahedrons. Their special affinity is for cells in the human upper respiratory tract—in fact, for a specific protein molecule called ICAM-1 that resides on the surfaces of those cells. Cold viruses are not all-powerful: if you swallowed them whole—and quickly—you'd likely escape scot-free. Rub them on your arms—nothing. But inhale them, or touch your virus-infested fingers to your mouth or nose, and you are playing with fire. Some 95 percent of people exposed to a variety of rhinoviruses that they have never before come in contact with will be infected.

Rhinoviruses limit their activity to the respiratory tract because they become inactive at temperatures above 33 degrees Celsius. The lungs and stomach, which are at 37 degrees, are too warm. Within hours, the cells that are infected begin churning out new viruses, the immune system kicks in on your behalf (but, unfortunately, is responsible for most of the unpleasant symptoms) and eventually, days later, it's all over. At least for you, but the chances are good that some of the millions of rhinoviruses hatched in your body seeded new infections in others.

So rhinoviruses are pretty simple. But that's not the case with all viruses. For instance, HIV, the human immunodeficiency virus, rather than switching into high gear once having gained access to a cell, may opt instead for transporting its genes right into the nucleus of the cell and leaving them there, undetected, for cycle after cycle of cell division. On the other hand, if activated (a process that is still mysterious), multiplication will begin. Multiplication finishes when the new virus particles start to exit the cell by traveling all the way back to just inside the cell membrane, then moving through and budding off, wrapping themselves in a little piece of the host/victim cell membrane as they go. That cell membrane is familiar and therefore not alarming to the immune system—unfortunately.

HIV is a sloppy reproduction machine, incorporating mistake after mistake into the new copies of its genes. This sloppiness actually makes it much more difficult for humans to invent ways of stopping the virus in its tracks. Every mutation in a gene inevitably creates a difference in the structure of the protein coded by it. Every such protein difference results in a virus that is just slightly changed and so slightly less recognizable, either by the immune system or by drugs. As is the case in evolution, most mutations are deleterious, but some help, and in the case of HIV there have been many such helpful mistakes.

There are some humans—about 1 to 2 percent of Caucasians of northern European descent—who are lucky enough to have a mutation in the gene that codes for one of the two cell surface receptors that will recognize HIV (or vice versa). In their case, this receptor never makes it to the surface of the cell. The cell is therefore invisible and inaccessible to the virus, and such people are almost always immune to HIV.

It's not surprising that HIV, a relatively recently discovered virus, is still mysterious. But so is the influenza virus, even though it has been intensively studied for decades. We still don't really know what to expect from the flu. We were recently

concerned about the possible evolution of avian flu into a viru-
lently human version, but at the time of this writing that has
not happened. However, we *did* have a kind of underwhelming
pandemic of H1N1.

Even when it comes to one of the most-studied influenza
viruses of all, the strain that caused the pandemic of 1918, it's
still not exactly clear why it killed more people than were killed
in World War I (the low estimate is forty million, but the figure of one hundred million is sometimes offered). That flu was
unusual because most of the fatalities were among fifteen- to
thirty-five-year-olds, a segment of the population that isn't
usually bothered by the regular annual flu. And they were
dying hideously, their lungs filled with fluid and blood.

A couple of years ago, enough 1918 viral RNA had been
found in formalin-preserved tissue samples in Washington,
DC, and remains from graves in the permafrost in Alaska to
allow researchers to piece together the exact details of every
one of the eight genes from that virus. In a move that some bit-
terly criticized for being too risky, one team assembled the entire 1918 genome—in effect re-creating the deadly 1918 virus—
and infected mice with it. The mice showed some of the same
extreme symptoms that the human victims showed nearly a
hundred years ago: a flu that came on with extreme sudden-
ness and hit the lungs particularly hard.

Experiments have suggested that a single gene is responsible for the exaggerated damage, but some scientists think it
looks like the eight genes in concert were somehow responsible for its deadliness. One thing seems to be sure: the 1918
flu virus triggered an over-the-top immune response, rallying

so many chemicals and so many kinds of cells to the defense that they damaged the body they were supposed to be protecting, the viral version of a combination of collateral damage and friendly fire. Even the experts grope for words that can adequately describe the violence of this process while remaining true to the prose standards of the research world. One such description is the "cytokine storm," cytokines being ubiquitous chemical messengers.[1]

You'd think, with only eight genes to deal with, it would be easier to understand the virulence of this virus. But it's not, and one of the reasons that we're just finding out about this upregulation of chemical signals that attract other cells (and also set in motion self-destructive processes) is that unraveling or even discovering all the signaling systems on a single cell is just too complicated.

The influenza viruses have a high rate of mutation, high enough that it is hard for us to keep up with the new varieties. Every flu season there's a slightly different human virus (again the result of subtle chemical changes produced by small mutations in the viral genome), but usually a flu vaccine can be made that anticipates roughly what that new virus will be like. Even so, the flu shots aren't very effective in preventing the illness. They offer at best partial protection, but even that is beneficial to seniors, the segment of the population that is usually the most susceptible to the flu.

Research into the much deadlier 1918 variety has shown that its immediate ancestor was an avian flu virus, but that in itself is not surprising, since most of the influenzas we endure have their roots in birds. What was different about the 1918 virus was that it appears not to have arisen the way most pandemic human viruses do. Usually two viruses simultaneously infect an animal and mix their genes into a unique and more lethal combination. The 1918 virus didn't do that but instead likely spent some years incubating in a host that was neither bird nor human. No one knows what that host was.

In many ways, we're still in the dark about the flu. But one thing is sure: the emergence of new strains, the ability of those strains to infect humans and the deadliness of that infection are all ultimately determined by the arrays of molecules on both the aggressor, the virus, and the target, our cells.

Where rhinoviruses stand out as exemplars of stripped-down viruses, in the same way that the dune buggy is a minimalist's car, both HIV and influenza represent the complex end of the spectrum of viruses. They have envelopes, surface spikes and probes; a library of tricks to use once they're inside their target cells; and the ability to shape-shift to avoid immune defenses and drugs. But that complexity doesn't necessarily correlate to virulence, and Peter Medawar's memorable description still holds: viruses are essentially nothing more than bad news (genes) wrapped in protein. For the thread of this story, it is the viruses that satisfy this minimalist definition that are the relevant ones. Many are less than a fifth the size of HIV and influenza, with half as many genes, just enough stuff to be able to gain access to cells, take them over and replicate. These viruses measure something like twenty nanometers, where a human hair is eighty thousand.

It was simple viruses like this that were in the backs of researchers' minds in the early 1960s. Scrapie, kuru and CJD surely weren't bacterial diseases, but they were infectious (despite that nagging absence of the usual infectious responses of fever and inflammation), and the only thing that seemed likely to fit the bill was some sort of virus, albeit one that took its sweet time to display disease symptoms. Back in 1959, when Bill Hadlow first realized the similarities between kuru and the sheep disease scrapie, and at the same time inspired the first experiments to show that kuru and CJD could be transmitted, you might have expected that the human diseases would get most of the attention of researchers. But that was not the case. It is a lot easier to follow up on whatever good, in-depth, inventive science has already been done than it is to start out blindly. Scrapie had been much more closely studied than either CJD or kuru, and Hadlow's realization that scrapie and kuru seemed to have a lot in common sealed the deal.

What was known about scrapie would apply directly to the other two diseases, and, whatever sort of peculiar virus it would turn out to be, concentrating on scrapie would give scientists a head start. Well, you would have thought so, anyway. As it turned out, uncovering the truth about scrapie just deepened the mystery.

7

When Is a Virus Not a Virus?
When a Disease-Causing Agent Reproduces Without Genes

The physicist I. I. Rabi uttered one of the greatest comments ever in physics when, presented in 1937 with the unexpected discovery of the elementary particle called the muon, the first particle discovered that does not exist in ordinary atoms, he said, "Who ordered that?"[1] But had Rabi been a biochemist or virologist in the 1960s, he could just as well have been talking about the mysterious agent that caused scrapie: it definitely was not on the menu.

As soon as it was established in the 1930s that scrapie could be passed from sheep to sheep, as soon as *infection* was accepted as the method of transmission, people started to scratch around in all the old familiar places. If it's an infection, they reasoned, there's an agent that causes it, and that agent is likely a bacterium, fungus or virus, so let's figure out what it is. Actually, not a bacterium—in this ultramicro world, bacteria are beasts, easily visible even with a desktop microscope, but they were absent. Eliminate bacteria and thoughts naturally drift to viruses, but unfortunately scrapie, like kuru, appeared to produce none of the usual signs of viral infection, like inflam-

mation and fever, that signal that an infectious agent is being
fought by the immune system.

There was also the incubation problem: it took forever for
these "viruses" to take hold after they were inoculated. Rather
than symptoms appearing after days, it was months or even
years. If they were viruses, they had to be "slow" viruses, a label
that had been officially adopted to describe a cluster of agents
that were, well, slow. But even though you could put a label on
scrapie if you had to, by no means did anyone think the disease
was perfectly understood.

The first person to closely examine the agent was a little-
known English scientist named David Wilson. He established
quickly that it was indeed small: at least as small as a virus.
He also put some boundaries on that prolonged incubation: it
took anywhere from eighteen to sixty months after inoculation
before the symptoms of the disease appeared in sheep. As well,
he found that this thing, whatever it was, was one tough nut.
It resisted destruction by a wide range of powerful techniques,
including formalin, phenol and chloroform (all chemicals that
would destroy normal viruses). It was still infectious after having spent two years in dried brain tissue; it survived half an
hour at 100 degrees Celsius, and it shrugged off deadly doses
of ultraviolet light.

Wilson, in direct contrast to the larger-than-life Nobel
Prize–winning characters in this saga, published next to nothing about his own work. It's been suggested that his results
seemed so weird to him that he became reluctant to make
them public. (This, of course, runs counter to the habits of
those who, if they encounter "weird" results, will *rush* to pub-

lication.) By all accounts he was shy, straightforward and honest, a careful and methodical scientist, slave to the inconveniences of the disease he was studying. Look at it this way: if he had been able to infect animals that reproduced with lightning speed, like lab mice, there might have been the chance that they would show symptoms within weeks, and he would have been able to do experiment after experiment in the time it actually took him to do one. He had to inoculate sheep, then wait several months for the disease to take. But he did exactly that, *nine* consecutive times, from 1945 to 1950.

Not only was Wilson the forgotten man because he published so little in the scientific literature, but, according to Bill Hadlow, by the end of the 1950s he had become a recluse.[2] Another researcher revealed that Wilson had had a nervous breakdown "because the Agricultural Research Council was extremely critical that he hadn't produced a vaccine yet."[3] Produce a vaccine? He was just finding out what the thing was! It's a rare scientific paper that reveals even a trace of emotion, but in a paper published in 1959 about scrapie, the bibliography makes an unintended but telling comment about Wilson, citing his research in the following way: "Wilson, D. R. Unpublished work. (1952); Unpublished work. (1954); Unpublished work. (1955)"[4]

Anyone who had been closely following the scrapie saga would have found Wilson's findings unexpected, peculiar and maybe even unbelievable. But Wilson's work wasn't going to make any impact until someone else corroborated it.

As interest in scrapie ramped up during the 1960s, a radiation specialist, a feisty woman named Tikvah Alper, started her

own efforts, together with specialists in the disease, to characterize the scrapie agent.[5] Alper knew all about radiation and
was very familiar with the effects it could have on living things.
She was working on a way of figuring out the size of very small
infectious agents, like bacteria and viruses, by calculating how
much radiation was needed to inactivate them. The bigger the
target, the more radiation was required. She bombarded samples of scrapie-infested tissue, then inoculated mice, not sheep,
to see if they'd get scrapie. Above a certain level of radiation,
the scrapie would be inactivated, the inoculated mice would
live happily ever after and, from the dose she'd used, Alper
would be able to tell how big the scrapie agent was.

This new approach confirmed that the scrapie agent wasn't
very big, about the size of the smallest virus known at the time.
That didn't prove it was unique or even particularly sensational, but the fact that its size was confirmed as being at the
extreme end of the scale was intriguing. Alper then switched
approaches and flooded the scrapie agent with ultraviolet light.
By then it was well known that UV light, especially at certain
frequencies, destroyed DNA, the genetic material of all bacteria and many viruses, or the very similar RNA, which served
for the rest of the viruses. In other words, if the thing causing
scrapie was *any* kind of virus, UV light would inactivate it.

But it didn't. The scrapie agent proved enormously resistant
to UV light. Unlike Wilson before her, Alper was not at all
reluctant to air her views about scrapie. She published an article in 1966 in which she argued that maybe the scrapie agent
wasn't inactivated by UV light because it could reproduce
without having either DNA or RNA, or as she put it, "the agent

may be able to *increase in quantity*," which presumably isn't quite the same as saying that it was reproducing.[6]

That was nothing less than heresy. Infectious agents had to have their own genes, whether DNA or RNA; the idea wasn't even up for discussion. It has been suggested that, well aware of the reception she'd get, Alper hedged her bets a little by publishing in a second-rank journal, but the bottom line is she published.[7]

Though she could have been forgiven for waiting for the dust to settle, Alper went on to refine her experiments by being much more selective about the exact wavelength of UV light she was using. She was then able to compare the effects of UV light at the precise wavelengths known to disrupt DNA and RNA with UV a little further along the spectrum, where it was known to disrupt proteins. She got what she must have been expecting: UV doses known to break DNA apart had no discernible effect on scrapie, and even the doses at protein-destroying wavelengths didn't do much. But she had been able to confirm her first, extraordinary claim that neither DNA nor RNA seemed to be involved in the reproduction of the scrapie agent.

This time around she threw caution to the wind and submitted her results to that giant of scientific journals, *Nature*. And she was turned down, at least initially. Angry protests finally persuaded the editor of the journal to publish her paper, titled "Does the Agent of Scrapie Replicate Without Nucleic Acid?"

Alper was pretty thick-skinned, and she had to be, because most biologists must have dismissed her suggestion as impossible. They had good reasons for believing so—the mid-1960s was the golden age of molecular biology, and scientists every-where had come to understand just how genes worked and how reproduction wasn't possible without them. To say Alper had committed heresy isn't an overstatement because the idea that genes were at the heart of all reproduction was enshrined in something called the "central dogma."

The phrase was coined by Francis Crick, of Watson and Crick and DNA fame. (Funnily enough, Crick, who was an incredibly smart man, was careless in his choice of the word "dogma." He had intended it to mean something grander than just a hypothesis, something for which there was no experimental evidence. Because he had no interest at all in religion, he ignored the obvious religious connotation of dogma as a belief that cannot be challenged and used it instead to mean something that could have far-reaching effects, despite the lack of evidence.) It was a catchy way of making the point that information in a cell is a one-way street. It starts with the DNA in the genome, which is transcribed into a short-lived copy made of RNA, which in turn is a template for the manufacture of a protein. "DNA makes RNA makes protein" is the way many high school students learned it. Crick liked to say, "Once [sequential] 'information' has passed into protein, *it cannot get out again.*"[8]

It made, and still makes, perfect sense. Cells evolved to reproduce. In the course of that evolution, machinery was assembled to translate the DNA code into the production of proteins, which then go where they have to go and do what they have to do. It is a damn complicated process, requiring lots of energy and the production and action of a host of dedicated molecules, all of whose job is to manage that *one-way* process, DNA into protein. Cells can't divide (and therefore can't multiply) unless they grow. They can't grow unless the requisite proteins are manufactured, and that can't happen unless the DNA is translated.

Nature has no need—and therefore provides none of the mechanisms necessary—to run that process backward. It would be like taking your car, returning it to the factory and requesting that it be turned back into its component parts. And not simply taking it apart but having the bumpers and engine parts and the fabric from the seats converted back into the design software that produced them.

While you might have learned the central dogma as "DNA makes RNA makes protein," its original description, by Crick himself in 1958, left the issue more open than that. He allowed for the possibility that RNA could make copies of itself (which indeed it can) and even that RNA might make DNA copies of itself (the reverse process of the normal, which some viruses actually do). But even in 1958, Crick was adamant that there was no foreseeable way that a protein could reverse the process and make the DNA that corresponded to it. So Alper, by showing something similar (that a protein-only agent could reproduce), was challenging biologists who were still flushed with excitement over DNA, the genetic code and the central dogma.

It isn't known whether Alper and Crick ever met face to face, but if only they had. They were, despite a difference in height of probably a foot or more, two peas in a pod. Both were great scientists, both were known for their incisive comments and neither had any patience for sloppy arguments, which, when tied to those incisive comments, apparently made many of their colleagues uncomfortable.

But even though Alper had challenged Crick's central dogma, her challenge had little chance of making any headway. Crick was one of the founders of the new molecular biology, a canvas being painted with dramatic speed, and the curiosity that was scrapie had to stay in the background, although Crick himself was not unaware of it. In 1970, he defended the central dogma but allowed that it was still much too early to say that it was correct: "There is, for example, the problem of the chemical nature of the agent of the disease scrapie."[9]

It might all sound a little parochial, but for biology Alper's assertion was the equivalent of challenging Einstein's theory of relativity, or the nature of the atom. If things multiply, they had to have genes to direct that multiplication. So it wasn't surprising that the vast majority of biologists either ignored this awkward contradiction or just assumed that with time it would go away.

But it was the 1960s, after all. Though we are talking about middle-aged English scientists with skinny ties, the small number of adventurous among them, perhaps sensing that something radical might be happening here, tried their hand at inventing explanations for scrapie. At one extreme was a scheme for allowing the agent to reproduce in the complete absence of genetic material, thus complying with Tikvah Alper's results. It was put forward in 1967 by the mathematician John Griffith. Sensitive to the fact that Alper's evidence of an agent-without-genes was causing distress in the biological establishment,

Griffith took pains to show that a scheme could be worked out that would allow proteins to reproduce *all by themselves* without seriously challenging the central dogma.

Griffith's treatment of the problem was a little bit dry, a set of equations and diagrams focused on proteins with different conformations and the energy required to shift them from one to the other. He suggested that by introducing some sort of unconventional protein into a vast population of its conventional counterpart, the renegade version could somehow alter the others, recruiting them to unconventionality; the tide would then turn, and the unconventional version could grow its numbers dramatically. It might not be reproduction in the traditional sense, but it *was* reproduction.

Griffith had hit on something, but it wouldn't be fully realized for another fifteen years, and his work somehow seems to stand off to the side, alone. It is always referenced, but in a way that seems as if academics feel they must, even if they don't think he really influenced the way things went.[10]

They didn't pay a lot of attention perhaps partly because his was a mathematical approach, partly because so much was still invested in the idea that infectious agents just had to have genes. In fact, research suggested that there were different strains of scrapie, slightly different versions that caused different symptoms, and it seemed unbelievable that this could be possible without the existence of slightly different sets of genes in the infecting *thing*, whatever it was.

Besides Griffith, who seemed clearly out on a limb, there were others who reined in their radicalism somewhat to come up with visions of scrapie that were more in tune with the biology of the time, though still outside the mainstream. It was a time for shrugging off the old ideas about infection and, if you had a theory, tossing it in the ring. Here was a disease, scrapie, that became more and more puzzling with every passing experiment. But the beauty of it was, even if you had never seen a sheep with the disease (one could go further and say even if you'd never seen a sheep, period, but that would be unlikely in England), you could still offer your thoughts. It was all about molecules, not organisms.

So there were those who thought it was a virus, one of the slow viruses, definitely peculiar, but a virus nonetheless. Others, who shied away from the virus idea—this thing was just too weird—but weren't willing to shy away completely, called the scrapie agent a viroid, or even a virino. Viroids are small pieces of naked RNA that infect plants. Somehow they are able to survive in the absence of the protective protein coat that seems so crucial to viruses. But one of the major problems facing the viroid theory was that the scrapie agent survived treatment that would kill DNA and RNA.

The virino was something else. A couple of the most prominent scrapie researchers in Edinburgh came up with this hybrid concept. They were traditionalists in the sense that they didn't buy the idea that the scrapie agent was a mere protein molecule. They were convinced that genes were somehow involved, but at the same time they were well aware that there was something unconventional happening. So to account for the strange properties, they invented the virino.[11] It was presumed to be a short piece of DNA that wrapped itself in a protein made not by its own genes but by the host. With that protection it could spread from cell to cell without risking destruction by the immune system, because immune cells wouldn't see it as foreign. The fact that it wouldn't have to manufacture its own coat meant that it needed only to maintain a minimal genome, just a gene or two, maybe just enough to take over control of some of the host DNA and do its thing. In fact, something like a virino, which is set apart from the host cell by only a tiny amount of DNA, might never be discovered.

It wasn't just virinos and viroids. The carnival of ideas about scrapie cast the mysterious agent as a piece of cell membrane, or some sort of elaborate sugar molecule, in addition to the persistent idea that it could be a protein. Intense rivalries formed, and the heat was on to figure it out. And there's no doubt that the heat had been turned up by Hadlow's discovery that scrapie was just like kuru.

But remember, kuru had been linked not just to scrapie but also to the intermittent and rare human disease Creutzfeldt-Jakob. And it wasn't long before it too provided researchers with food for thought, although in a much more frightening way than scrapie.

8

Creutzfeldt-Jakob Disease:
Waking Up to the Potential
of a Devastating Affliction

Insidious is the word for scrapie, kuru and Creutzfeldt-Jakob disease. In the case of kuru, it was the subtle but inexorable attack on the brain without any of the usual signs of disease. With scrapie, researchers were baffled by the agent itself, something too small and seemingly too simple to be able to do what it did. And CJD? By the late 1960s, Carleton Gajdusek and his colleagues in Washington had shown that CJD inoculated into animals produced symptoms consistent with the human disease, but—not surprisingly—only after years of waiting. Yet, at the same time, unlike kuru, there was no evidence that anyone in the general population "caught" CJD. In one sense there still isn't any evidence for that, but it became clear in the 1960s and 1970s that this statement isn't completely true: it really depends on what you mean by the "general" population. There are people who do catch CJD, but only if they find themselves in certain special, and unfortunate, circumstances. This group of victims testifies to the incredible hardiness and opportunistic nature of the CJD agent, the same thing that scrapie researchers had found with theirs.

The rough outlines of Creutzfeldt-Jakob disease were already clear by the 1970s. It has pretty much the same incidence everywhere in the world: about one person in a million every year comes down with the disease. And when they do, they don't last long. Most victims are dead within a year, and their decline is not a pleasant one, starting with relatively mild symptoms of depression and confusion but progressing to shaking limbs, dementia, blindness and death. There isn't much time to prepare, and, perhaps fortunately, there isn't much time to contemplate what's happening. Reading the accounts of these deaths, as clinical and dispassionate as they are in the scientific literature, is difficult. The disease affects men and women equally and usually strikes people who are in their late fifties or early sixties. (Curiously, the likelihood of coming down with CJD declines after the age of seventy, but why that is, nobody knows.)

In other words, it looked sporadic and random. There did seem to be the odd cluster here and there, a handful of cases appearing in a relatively short time within kilometers of each other. But these clusters were rare, and an overall rate of one in a million doesn't mean that all the cases will be evenly spaced geographically and over time. While the disease appeared to be sporadic, something surely *did* cause it. But why would it attack one person rather than another? Surely it couldn't be as accidental as a lightning strike. How did you acquire the agent? Of course, all those completely reasonable questions were shadowed by the specter of kuru. Kuru was infectious: maybe so too was CJD.

As the 1970s passed, it became clear that the agent that

caused CJD was everything researchers had feared and more. Yes, it was indeed sporadic. Yes, most victims seemed to have no connection with each other. But on the other hand, CJD found ways to infect people that had never been anticipated.

For instance, in 1974, a group of physicians at Columbia University in New York City reported that a woman who had received a corneal implant from a man who was later found to have died from CJD came down with the disease herself eighteen months after the implant, endured the same distressing list of symptoms and died with more extensive damage to her brain than even had been suffered by the donor.

There have been three more cases of corneal transplants being responsible for the transmission of CJD. One of those patients had her transplant in 1965 but didn't develop CJD symptoms for another thirty years. That is an extreme incubation period—the other cases of corneal infection usually showed up within a year and a half of transplantation.

Two years later, in 1976, doctors in Switzerland reported the cases of two CJD victims apparently infected by silver electrodes that had been temporarily implanted in their brains. The source of the infection was a woman in her late sixties who, through the summer and fall of 1974, had been going downhill with the classic CJD symptoms of depression, anxiety, memory and speech difficulties, vertigo, jerking limbs and mental deterioration. In September of that year, as part of a surgical attempt to quell the violent lashing about of her legs, doctors implanted electrodes in her brain for two days (by this time she was unresponsive). She died three months later.

Meanwhile, a young woman in the same hospital was being

CREUTZFELDT-JAKOB DISEASE

treated for epilepsy that was too severe to be controlled by drugs. As part of her treatment she had recordings taken from inside her brain in the areas where the seizures were thought to be triggered. Nine electrodes were used, and two of the nine had been implanted, nearly two months before, in the brain of the CJD patient. Six months later, the young woman noticed she was having trouble walking, her memory was impaired and she couldn't speak properly. Her EEG showed clear signs of CJD.

But the two electrodes weren't finished. A month after the young woman was infected, they were part of a set of seven that were implanted into the brain of a young epileptic—he was seventeen at the time. Two years later, in April 1976, he began to experience the first symptoms of CJD and died before the end of that year.

The electrodes used on these patients were sensitive to heat, so they hadn't been sterilized that way; instead, they had been disinfected with alcohol and "sterilized" in formaldehyde vapor for forty-eight hours. There had never been an infection before—this was a tragic first.[1]

If you look at just these two cases, you have to suspect strongly that anything coming into contact with a CJD-infected brain is going to become infectious itself. Of course, it's a lot easier to see that now than it was to have figured it out then. Nevertheless, one person did. In the mid-1970s, researcher Alan Dickinson was immersed in scrapie research in the United Kingdom, but he had a strong interest in CJD too, and tells the story of how one night he had a terrible thought about human growth hormone. At the time, it was being harvested from the pituitary glands of the recently deceased, then injected into young children whose growth and, just as important, well-being were compromised by a lack of that hormone. It was effective too: daily injections could take a child from growing half as fast as others to *twice* as fast, within a year. But Dickinson feared a disaster: if even one of those pituitaries was from a patient who had died of CJD, many young children would be at risk for the disease.[2]

Dickinson was tragically correct: the techniques being used did not exclude the CJD agent, and a tragic new cycle of inadvertent infection, sometimes called therapeutic misadventure or iatrogenic infection, began.

Human growth hormone treatment was hugely popular; it has been estimated that something like two hundred thousand children worldwide received it. But when the math was done to calculate the risk that a CJD-infected pituitary could get into the mix, the results were a little scary. Even though CJD is one in a million, autopsies that focused on the brain (and so would have made the pituitary available) were usually performed on those who had died of neurological disease, dramatically raising the chances that CJD might have been present. One estimate suggested that one pituitary in a thousand might have been infected. Hundreds of thousands of glands were processed, meaning that *hundreds* of infected glands could have entered the system. No one can confirm that number, but Dickinson's fears were realized: two hundred people died as a result of having tainted hormone injected into them.[3]

There were even further cases of accidental infection. Surgeons commonly used pieces of donated dura mater (the membranous covering of the brain) as a patch to help cover and protect the brain after an operation. Maybe it was inevitable, maybe not, but one particular batch of dura mater produced by a German company and used around the world turned out to be contaminated with CJD. Having such material plastered directly onto the brain was a golden opportunity for infection, and to date another two hundred cases of CJD have been caused this way. The picture once again highlights the length of the incubation period: even though use of this particular batch of dura mater was stopped in the 1980s, cases continued to rise through the 1990s, peaked in the last couple of years of that decade and only started to diminish into the 2000s. There will likely be new records still to be set for length of incubation period, but so far it is a whopping twenty-three years for dura mater transplants and thirty-six years for human growth hormone. It's not really clear why, in these particular situations, the incubation period is so long.

Somewhat encouraging is the fact that no new cases of surgical instrument contamination have appeared, even though it is argued that the sterilization procedures are still, as the authors of a recent article noted, "suboptimal." They suggest that it is improved screening of donors that is keeping us safe.

Imagine being back in the 1970s in the middle of this rising awareness—and apprehension—that these infectious agents could get a toehold in the unlikeliest situations. As always, with so little information available, it was extremely difficult to distinguish the case that could start a miniepidemic from one that, like so many, signified nothing. For instance, a series of CJD patients was reported back in 1960, several of whom had had brain surgery prior to their diseases, for reasons completely unrelated to their disease. Did having your head opened, under certain conditions, predispose you to CJD? And if it did, what exactly were those certain conditions? Few now think there's anything significant going on there.

There has also been a handful of medical personnel—a neurosurgeon, a neuropathologist and two medical technicians—who have come down with CJD, all of whom might have come in contact with infected neural tissue. Are those professions more hazardous than we thought? Counting against that idea is the fact that the rate of CJD among these professions is no different from what it is in the general public.

No trip through the 1970s, the decade that woke everyone up to the potential of CJD, would be complete without the story of the Libyan Jews. Almost the entire Jewish community in Libya, probably thirty thousand strong, immigrated to Israel between 1949 and 1951. The story of CJD in those people started with a routine study of half a dozen cases in Israel in 1972. Of these six, three had emigrated from Libya. The authors of that study also reported a second set of six, not their subjects, in which two were from Libya. That's a total of five out of twelve, which, the authors said, was "of some interest." It was a passing comment only; the paper was devoted to analyzing similarities and differences among the symptoms.

Five cases isn't a lot, but percentage-wise it was enough to trigger, if not alarm bells, at least intense curiosity. A subsequent nationwide survey in Israel turned up twenty-nine CJD cases, a mix of certain, probable and possible, and still the

Libyan factor stood out. The rate of CJD in Israel among non-Libyans was the standard worldwide figure of around one per million. In Libyans, it was thirty-one per million. The male-to-female ratio was a little higher too, and the duration of the illness shorter, but these details were insignificant beside the enormous susceptibility of the Libyan immigrants. As mentioned, the shadow of kuru hung over every CJD investigation in the 1970s, but especially in this case. At the end of the report the authors referred to kuru specifically, arguing that because exposure to human nervous tissue was crucial in that setting, "parallel studies in Libya would be desirable."[4]

Did scientists have cannibalism in the back of their minds? Not exactly, but they were wondering about something very much like it. A report shortly afterward—by a group in Bethesda, Maryland, that included Carleton Gajdusek—argued that it wasn't brains in this case, but *eyeballs*. Sheep eyeballs. These scientists put three things together: one, that sheep eyeballs were a delicacy among a variety of North African peoples; two, that scrapie was a sheep disease not unlike CJD; and three, that a recent case of CJD had apparently been caused by an infected corneal transplant (as we've seen).[5] This suggestion met with a somewhat caustic reply from Milton Alter, one of the authors of the Israeli study. He kicked off by pointing out that the food habits of other cultures never failed to "amaze 'enlightened' Westerners," then added that while they were right that eyeballs were consumed (and only lightly grilled, thus allowing possible infection), it wasn't just CJD victims who ate them—virtually all North Africans did, and plenty of Europeans too. He also reminded readers that ingesting these agents was much less likely to cause infection than being inoculated (i.e., through corneal transplants) with them.

But Alter couldn't dismiss outright the possibility that consuming infected sheep eyeballs had something to do with the Libyan puzzle. And the need for an explanation intensified when Alter and Esther Kahana published a follow-up study of CJD in Israel that had turned up twelve more cases, eight of which were Libyan, pushing their rate much, much higher. In this case, the victims had all been in Israel for at least twenty-three years, but the notoriously long incubation period could allow for that. However, Alter and Kahana now strengthened Alter's earlier argument that there seemed to be no difference between the eating habits of Libyans with CJD and those without. They were able to establish that at least one-third of patients had eaten sheep eyes and *brains,* and yes, they were often wrapped in paper and grilled for only five minutes or less (not enough to kill CJD), but so had a *larger* number of people in the healthy control group. It was also true that there was no scrapie in Libya and that many other Jewish communities around the Mediterranean ate sheep brain yet had no elevated risk of CJD. So culinary habits seemed a poor explanation for the astonishingly high rate of CJD among Libyan Jews.

So what was going on? It really wasn't until doctors identified a CJD patient of Libyan origin who had been born in Israel and had never consumed a sheep's brain that attention turned to genetics. It's easy to see how this explanation might have been overlooked at first: there wasn't much accurate information on cases within families prior to their arrival in Israel, people were loath to talk about dementia running in their family, and death from other causes likely killed some CJD carriers before they began to show symptoms.

But genes did turn out to be the problem: most of the Libyan Jews carried a gene that made them not just prone to CJD but carriers of it. History accounts for the fact that this gene is so widespread in that population. Some accounts of the settlement of Jews in Libya say it began about nineteen hundred years ago, after the destruction of the Second Temple in Jerusalem by the Romans, but others contend that the first Jews arrived in that part of the Mediterranean much earlier, centuries before the Roman Empire even began. One account puts the Jews arriving on the island of Djerba (now in Tunisia) around 600 BC. Partly because of persecution, partly to avoid assimilation, they remained separate and intermarried heavily, with the result that the CJD gene spread thoroughly through the community.

The Libyan story put genetics in the picture for the first time, establishing that there were really three kinds of CJD: sporadic, inadvertent (iatrogenic) and familial. The puzzle of CJD was gradually coming together, while at the same time getting more and more complicated. But that should have come as no surprise—look at what was happening with scrapie. The more research was done on both, the more exotic the causative agent became. A breakthrough was needed. There was one on the horizon, but it wasn't the kind anyone could have anticipated. And that breakthrough can't be appreciated in all its glorious detail without a brief detour through the world of proteins. That's next.

9

Magnificent Molecules
The Proteins That Make
Life Possible

In his satirical 1963 science fiction novel *Cat's Cradle,* American writer Kurt Vonnegut relates this short but telling exchange:

> "What is the secret of life?" I asked. "I forget," said Sandra. "Protein," the bartender declared. "They found out something about protein." "Yeah," said Sandra, "that's it."

Proteins *are* the secret of life, and the molecular biology revolution of the 1960s was pulling scientists deeper and deeper into their world. Proteins, and how they are built, were on everyone's mind, especially the researchers chasing the agents of scrapie. Those mysterious agents were controversial simply because they appeared to *be* protein—and nothing else. And while esoteric schemes had been proposed to show how such protein-only objects could reproduce, few believed in them, or were even aware of them. But even if you were a skeptic, unwilling to accept the idea of reproduction in the absence of RNA or DNA, unable to envision a form of life made of protein only, you still had to admit that protein had a very, very important role in scrapie and its sibling human diseases.

Proteins are magnificent molecules, complex but subtle, possessed of an architecture that allows them to make life—and death—possible. Hold a silk shirt in one hand and a wool sweater in the other, close your eyes and feel the differences between the two fabrics. Rub them between your thumb and forefinger; take one and pull on it to see how much it will, or will not, stretch. As you do this you are performing a molecular experiment; the qualities of silk and wool, even though both are proteins, are due to the constituents out of which they're built and the way they're assembled. The truly amazing thing is that the differences between them, which exist on a level that is difficult to visualize even with the most sophisticated technology, are so obvious to the relatively crude sense of human touch.

Protein molecules are in some ways like pearl necklaces: long chains assembled of relatively similar objects strung together. But the analogy is superficial at best. There are twenty different kinds of "pearls" in a protein molecule, twenty different subunits called amino acids, from which the chain is assembled. Given that there are twenty to choose from, and that proteins can be almost any number of amino acids, the variety of proteins that result is, practically speaking, limitless. And complex, unlike a pearl necklace. For one thing, although all twenty amino acids share a basic skeletal structure, each is adorned with something extra that makes it different from all the rest. These are molecular add-ons, a short side chain attached here, a sulfur atom there. If it were possible to create a larger-than-life protein chain, as big as a pearl necklace reaching down to your shoes, it wouldn't take much to see that each "pearl" is different. This is one significant difference between pearls and amino acids, necklaces and proteins, but not the most important.

In real life, proteins never exist as loopy chains. They are full of tightly packed twists and turns, folded on themselves to create lumpy, globular shapes that might bear a superficial resemblance to that pearl necklace if you crumpled it in your hands into as small a package as possible. But crumpling implies the application of undisciplined force, and that's not what happens here: creating a stable molecule is what counts, and stability is determined by push and pull at the atomic level. Those differences among the twenty amino acids, the side chains with which they're decorated, influence the shaping of the entire chain. Some have attachments that repel water, so those will exert strong chemical forces to hide on the inside, as far away as possible from the watery surroundings of the living cell. That will cause the chain to fold in on itself, in the most stable manner possible. Others, being "hydrophilic," will do exactly the opposite, exposing themselves to the milieu. These two opposites are a powerful force for bending and twisting the chain, and the exact final conformation depends in part on where such amino acids are located on the chain. If you have a segment that is hydrophobic, then that entire piece has to find shelter somewhere inside the final folded form.

As well, some amino acids will be desirous of bonding with others further down the chain, creating a loop. Those bonds can be formed between amino acids that are virtually at opposite ends of the chain. When all of these mechanisms are taken together, a stable protein results. It has been suggested that, over evolutionary time, the "principle of minimal frustration" selects sequences of amino acids that favor the most efficient and chemically secure folding.

Back now to the wool sweater and the silk shirt. Wool and silk are both proteins, but the way they're put together is so dramatically different at scales of a trillionth of a meter that you can feel it when you hold them in your hands. Wool is largely arranged as helices, coils of amino acid chains winding around each other and bonded to other coils. These coils can be stretched, sometimes up to twice their natural length, but they will nevertheless spring back when released. On the other hand, they are not all that strong. Silk is a different story: it is extremely strong because it is arranged not in coils but in sheets that are extensively cross-linked for reinforcement. Silk doesn't stretch nearly as much as wool, nor does it have the same elastic response. Pull on a silk fiber and it will first resist, then break.

Wool is in the same family of proteins as skin, horns, nails and hooves. The differences among them are, again, down to the individual amino acids and how they're organized. As you stretch a wool fiber, weak bonds among some of the amino acids break, but stronger ones between the sulfur atoms in some of them hold fast and provide the impetus for the bounce back. The more sulfur, the harder the substance, so hooves are sulfurous, wool is not. And skin is in between.

The examples I've offered so far are just a tiny sample of the world of proteins. In the human body, there are likely in excess of a hundred thousand different proteins, some inertly structural, like fingernails and hair, some tools and conveyances, some—the enzymes—promoters of chemical reactions. But in every case, the shape that the complete protein molecule adopts, that crumpled, folded necklace, is all-important. And one more protein, this one also a familiar one, illustrates the importance of shape even better than the wool and silk. It is hemoglobin, the molecule that is packed into our red blood cells, the chemical that traps oxygen in the lungs, then carries it to the rest of the body.

The word "packed" is completely appropriate—a single red blood cell contains about 280 million molecules of hemoglobin. And there are 5 billion red blood cells in every milliliter of human blood. So when you donate a pint of blood, you're giving away something over one quintillion hemoglobin molecules.

Hemoglobin is actually four protein molecules in one, four different folded chains bound together. Each of the four contains an iron atom buried in the middle of the folds, and it's that atom that grabs an oxygen molecule in the lungs, then releases it where it's needed in the body. It does the same thing with carbon dioxide, only in reverse: picks it up in the tissues, where it's been released as a product of active metabolism, and deposits it back in the lungs, where it can be exhaled.

Breathing is all about hemoglobin. But what a protein it is! Four similar chains bound to each other, each one a set of eight helices zigging and zagging and winding around each other to produce a pocket for the iron atom embedded in its supporting structure (the heme group). There is precision here—this is not analogous to stuffing bubble wrap around something you want to protect from a careless courier. Every single amino acid is crucial. That statement is more meaningful when you realize that there are about 140 amino acids in each of the four hemoglobin chains, and changing just one of those can have a serious impact. That does indeed happen, and sickle-cell disease is a perfect example of the result.[1]

In sickle-cell disease there is a single amino acid substitution in two of the four hemoglobin chains, so two changes out of 560 amino acids. Rare as they are, these changes exert a huge influence on the molecule. By changing the shape of the hemoglobin molecule slightly, adjacent hemoglobins can bond to each other, and this becomes a runaway process, creating long, twisted chains of hemoglobin within the red blood cells. These deform the red blood cells from their normal disk form to a sickle shape.

On the one hand, these sickle blood cells make life extremely difficult for the malaria parasite, which spends part of its life cycle in those cells. Obviously, in areas of the world where malaria is endemic, this would be a very good thing. But it is a two-edged sword: those same malaria-resistant red blood cells don't do well at high altitude; in some circumstances, especially where the oxygen concentration is low, these misshapen blood cells clog small blood vessels and can be fatal. It's estimated that two hundred thousand children under age five die in Africa every year as a result of sickle-cell disease.

Without my going into the molecular details, you can see that there must be intrigue here. What happens to the shape of the hemoglobin molecule as a result of that single, one-out-of-a-hundred-and-forty substitution? And how can that mo-

lecular tweak—that *single* molecular tweak—have those dia-
metrically opposed results of malarial resistance and oxygen
deprivation? It underlines one of the fundamental facts about
proteins: the form of a protein is its function.

There is one thing missing from this brief snapshot of proteins: every protein molecule is born as a simple linear chain of
amino acids, but moments later (even as the chain is still being
assembled) the protein is folding up, assuming its final form.
I suggested earlier that such folding emerges more or less automatically from the sequence—if this chemical group is here,
it can attract that one over there—but it is more complicated
than that. Complicated but crucial. As proteins fold, this story
unfolds.

10

Protein Origami
Building the Gothic
Cathedrals of Life

This, the second of two chapters about the structure of proteins, paints a picture of just how complex and puzzling the world of those molecules is, in particular the mechanics of the way they fold into the precise shape that allows them to function. It might seem like esoterica, but this bit of protein science, right out there on the leading edge, is crucial to the story of the weird agents responsible for scrapie, kuru and Creutzfeldt-Jakob disease (and, as it turns out, many other diseases too). And so:

Here's a strange thing. Even though the complete amino acid sequence is known for thousands of proteins, and even though the exact shapes of hundreds, if not thousands, of proteins are already stored in data banks, exactly how the configuration of each of those proteins is determined by its amino acid sequence is almost a complete mystery.

If you were to give a biochemist the sequence of amino acids of any protein, you'd think he or she should be able to look at the order, check where they're placed relative to each other, take into account all the chemical gizmos and doodads displayed by each amino acid and then be able to tell you with great accuracy what shape that protein will ultimately take. But no biochemist can do that. It's not that there's some mysterious life force operating here—all the chemistry and physics that dictates folding is known. It's also true that in hindsight it all makes sense; that is, once the shape is known you can see exactly how it was built. But going in the other direction, looking at the sequence of amino acids and predicting the shape, has been one of the most intractable problems faced by molecular biologists.

The challenge is that there is more than one way to get from here to there—an awkward excess of ways, according to some. Levinthal's Paradox (named after the late physicist and protein expert Cyrus Levinthal) holds that there are too many alternative ways of folding, and that no protein could ever try them all on its way to the most stable configuration. The total number of folding possibilities for a protein of one hundred amino acids turns out to be devastatingly high: it has actually been cited as "zillions" in a scientific article, although the authors then hastened to be more quantitative and estimated it to be 2 to the 100th power. The paradox estimates that it would take a protein 10^{10} years to settle on the right configuration. Well, if things really worked this way, proteins would *never* fold. Obviously we're missing something.

Levinthal thought the answer might be that a protein doesn't have to try every possible fold to "feel" which is best. Maybe it follows a well-worn path to the right fold. But little evidence has been found to support that suggestion. Adding to this puzzle is the fact that when a protein starts to twist and turn as it moves from its initial unfolded birth to its mature form, it may experience intermediate shapes that are somehow left on the cutting-room floor. You just don't see these motifs in the final protein. That means you can't take the final product and sight back toward the original and assume that the trip was a straight line.

It is a dynamic process, moving at an incredible pace. A short piece of protein thirty amino acids long can wrap itself into a spiral in twenty nanoseconds, twenty billionths of a second, a span of time not appreciable by us who divide our lives up into tenths of seconds at best. Peter Kollman's computer group at the University of California, San Francisco, has used supercomputers to show that over that same sort of time span, an entire protein will start to twist, bend and collapse into a globular shape, but while doing so it's all over the place, expanding, contracting, wobbling and wiggling like a wild animal in a burlap sack. There are strange periods of quiescence too, like between 250 and 400 nanoseconds, when the molecule is virtually still. But the protein is soon on the move again. Reminiscent of the trial-and-error folding process that Levinthal found so unlikely, Kollman wonders if it is in some sense searching: "It's a tantalizing idea—that the mechanism of protein folding is to bounce around until it finds something close, and stay there for a period, but if it isn't good enough it eventually leaves and keeps searching. It might have ten of those quiet periods before it . . . reaches the final structure."[1]

You might think that this is a sloppy way to run a business. Once all the amino acids are in place, why doesn't the protein, by doing a couple of twists and flips, take the cleanest route to the most stable shape and then stay there? One reason may be that it's worth trying out more than one final shape that is close enough to perfect stability. It's tempting to think of protein folding as the inexorable search for and eventual discovery of the perfect shape, the conformation that puts every atom in exactly the right place to maximize stability. But that would be a little too perfect for a biological system that has evolved over hundreds of millions of years. Genes have changed over that time; the way their proteins fold must have changed as well. It's easy to imagine (but hard to demonstrate) that some proteins, as they fold, revisit the way their ancestors folded, and if that's true, there must be false starts, some of which would have been bypassed or discarded. Whether history is involved or not, it might just be that the "most stable" shape for a protein is not an absolute but a relative term, and that there are, at least for many proteins, several three-dimensional structures that could compete for the title of "most stable."

Regardless of the ambiguities, it's a remarkable ability to be able to fold into working shape in this dynamic way. And in the early 1960s, protein scientists discovered the following: denature a protein—that is, treat it with heat or chemicals that would tear apart the folded architecture—then remove the denaturing agent and put the protein back into a suitable medium, and it will spontaneously and accurately refold. But that doesn't always happen, otherwise you'd be able to change your mind about eating those boiled eggs in the morning, cool them off and put them back, de-cooked, into the carton.

It's now known that there is an entire family of helper proteins called chaperones, whose job is to help proteins achieve the correct folding arrangement. The cell is a busy, crowded place, and unless they are surrounded by protective chaperones, half-folded proteins might get deflected from the path by other molecules, or stick to other partially folded proteins. In that sense, "chaperone" isn't as accurate as "bodyguard." Most of them are built so that they have a central cavity into which the protein needing protection inserts itself. Not only do these molecules ensure that folding is complete and accurate but they also protect proteins against stress. If you're running a fever, you're making more of those chaperones known as heat shock proteins, which help keep your proteins together in the face of denaturing warmth. And note the small irony that chaperones themselves *are* proteins.

And finally, this protein folding issue is more than just an esoteric game being played by scientists to satisfy their curiosity. In the end, protein folding will hold the key to a fundamental understanding of the cell and life. In the short term, it has practical aspects. There are times when the inability of proteins to fold in industrial processes makes their production much more complicated and expensive. For instance, the pharmaceutical giant Eli Lilly manufactures human insulin. Insulin is a relatively small protein, just fifty-one amino acids long. But manufactured insulin won't fold into the correct shape. Instead, improperly folded chains clump together and must be treated with chemicals to unfold them, then taken through a series of reactions to fold them up properly. It takes huge amounts of time and money to do what nature does with ease.

Living cells come up with the best protein structures on

the fly; thousands upon thousands of them are made, folded and put to use every second in those cells. They are impressive pieces of work, but it wasn't a lack of appreciation of their complexity in the 1960s and 1970s that made so many researchers argue that, no matter how diverse proteins were, they couldn't be infectious agents all on their own. Everyone knew they were the Gothic cathedrals of molecules, but everyone also knew these molecules couldn't build *themselves*. The central dogma had already made it clear: you needed DNA to make proteins—proteins did not make other proteins.

So here is the issue: as diseases like scrapie, kuru and CJD progress, the number of infectious agents found in the brain multiplies many times over. But if those agents are just proteins, where are the genes that made them? Well, nowhere, as far as anyone could tell. It was a standoff. Then, in the early 1980s, the story took another dramatic leap.

11

Stanley Prusiner's Heresy
An Infectious Agent That's a Protein and Nothing But

Dr. Gabrielle Nevitt at the University of California, Davis, studies the order of birds known as Procellariiformes. They include familiar marine species like the albatrosses, and a whole lot of lesser-known birds. It is no wonder they're not well known: most of them nest in the extreme southern oceans, close to Antarctica.

They are a fascinating group, these Procellariiformes, or "tube-nose" birds. They wander ceaselessly over the oceans, sometimes traveling distances that are hard to believe, simply to find food to bring back to their young on the nests. And diverse: the albatrosses are the biggest, but most are much smaller, including some of the cutest of them all, small plump birds, blue-gray, smelling like warm earth, a smell so loved by those who research them they keep bags with the birds' scent on them in their offices. These fragrant little birds are called prions.

Before 1982, these tube-nose birds of the southern seas were the only prions on this earth. But in 1982, a paper published in the journal *Science* not only relegated the ornithological prions to the back pages but completely changed the world of CJD, kuru, scrapie and a soon-to-be-expanded list of diseases.[1]

The paper was titled "Novel Proteinaceous Infectious Particles Cause Scrapie."[2] It was written by an American scientist named Stanley Prusiner, and it turned the world of research into the causes of these neurodegenerative diseases inside out. Prusiner argued that the infectious agent of scrapie (and presumably of CJD and kuru too) was not a viroid, not a virino, not a slow virus, not a piece of membrane, nothing but a protein, as Tikvah Alper and David Wilson had suggested before him.

So if he was revisiting an already-established idea, why was Prusiner instantly controversial? Perhaps because he was much bolder than Wilson or Alper: he not only argued forcefully that there was no question the infectious agent was a protein but had the gall to give it a new name, one he'd coined himself, even though scientists had been studying this disease agent for four decades plus.

This new protein was to be called, from then on, a prion. Prusiner helpfully pointed out that it should be pronounced *pree-on*. Prions, Prusiner wrote, are "small *pro*teinaceous *in*fectious particles." *Proins,* if you follow his italics. But Prusiner (correctly) figured that *prion* was more euphonious—and likely also realized that it offered fewer stupid rhyming opportunities. Apparently he was quite proud of the word: in an interview with science writer Gary Taubes in *Discover* magazine in 1986, Prusiner said, "Prion is a terrific word. It's snappy. It's easy to pronounce. People like it. It isn't easy to come up with a good word in biology. One hell of a lot of bad words people introduce get thrown away."[3]

I'm sure ornithologist Gabrielle Nevitt and her prion-studying colleagues would wholeheartedly agree that it isn't easy to come up with a good word in biology. Especially one that isn't already in use. But some insignificant southern ocean birds weren't going to derail what was becoming the prion express. For one thing, Stan Prusiner was on his way to ruling the world of these diseases and their causative agents, now with their own name. A Nobel lay in his future. And the study of prions, the infectious agents, is now a global research effort. So it's worth checking out just what Stan Prusiner actually said in this much-cited paper in *Science.*

It's a long report—more than five thousand words—and much of it simply reviews what was already known. That review is detailed and methodical, and as Prusiner surveys the entire scientific landscape, he also narrows the focus, and with great skill. For instance, he lists all the kinds of agents that had been proposed—including the ones already mentioned, like membranes and virinos, plus a whole lot more—then, applying the knowledge that had been gathered to that point, throws most of them out. Is it some sort of complex sugar, a polysaccharide or a naked DNA or RNA? Nope—there's no doubt that a protein is involved. How about a conventional virus, a parasite or even very unconventional bacteria? No again—they're all too big.

Inexorably, the paper chews its way through the alternatives and finally settles on a protein, but not without some cagey qualifications. At this point in the game in the early 1980s, even though you get the sense, reading between the lines, that Prusiner is setting everyone up for an agent that is protein and nothing but, he can't really go out on that limb completely without running a serious risk of having it sawed off behind him. So he kept one hand on the tree trunk by saying, "The term 'prion' underscores the requirement of a protein for infection; *current knowledge does not allow exclusion of a small nucleic acid within the interior of the particle.*"[4] In other words, he's confident it's just a protein, but he's willing to allow the possibility that there might after all be nucleic acid, DNA or RNA, some genetic material, even though in his mind there's no evidence for it. This kind of qualification was prudent, as his protein-only model was, in his words, "clearly heretical."

Having tantalized his readers with talk of heresy, Prusiner pulls back slightly, reverts to his methodical style and asks the rhetorical question, how do prions replicate?—pointing out that if it's true that prions have no genes of their own, then there are really only a couple of possibilities. One is that they could somehow use genes already present in the host cell to direct their own multiplication. He presents the fairly scant evidence for this, including the evidence that there are some familial (and therefore genetic) forms of prion disease, like the CJD that plagued the Libyan Jews. But there is also a problem with that: such deleterious genes would have trouble escaping the paring knife of evolution. Genes that kill people have the obvious side effect of limiting reproduction, and if there are no offspring, then they (and, of course, the organisms containing them) have reached a dead end. (Although in this case, given the advanced age of, say, CJD patients, death would have followed reproduction.)

The crucial card Prusiner is playing is that it's possible that genes aren't involved at all. If that's the case—and this *was* heresy, no doubt—the only two ways that proteins alone would be capable of reproduction would violate Francis Crick's long-standing central dogma. Either they could, Prusiner suggests, reverse the whole process by having proteins make genes instead of genes making proteins, something that seemed incredibly unlikely. Or the protein could make other proteins directly. Even though mathematician John Griffith had shown at least in a theoretical way that this might be possible, the vast majority of biologists would have deemed that to be practically impossible.

Prusiner finishes off this article with three short paragraphs that leave no doubt as to where he's going. The first repeats the idea that prions might be proteins only—no genes. The second argues that there might be implications here for a stunning variety of human diseases, including Alzheimer's, multiple sclerosis, Parkinson's, amyotrophic lateral sclerosis (Lou Gehrig's disease), diabetes, rheumatoid arthritis, lupus and several cancers. And while the third reiterates the importance of further research, at least partly to address the question of whether prions are truly devoid of genes, Prusiner leaves the strong impression that he doesn't really think such genes will be found.

And that's it—only a bibliography of 120 references follows. In that same issue of *Science* are articles to which most readers no doubt paid much more attention. One was about President Ronald Reagan's supposedly defensive MX missiles, another was about the U.S. State Department's claim that it now had unequivocal evidence that the Soviets were shipping toxins to Laos, Cambodia and Afghanistan to be sprayed from planes against insurgents. The main evidence, isolated from vegetation, was small amounts of oily stuff called yellow rain that contained a fungal toxin. Some experts countered that yellow rain was really bee feces.

So there was plenty to distract the interested reader from "Novel Proteinaceous Infectious Particles Cause Scrapie." But some readers definitely took it all in, and many weren't exactly thrilled with what they read. In fact, the article generated a vast amount of attention and criticism, and it polarized the scientific community. It may have appeared nearly thirty years ago, but the aftershocks are still being felt.

12

An Infectious Idea
The Campaign for the
Minds of Researchers

Stanley Prusiner's move to name the disease agents "prions" and to dispense with not just old terminology like "ultra-slow viruses" but the very *idea* of viruses was a bombshell. There are researchers today who have not abandoned their resentment of Prusiner and disdain for his behavior. In their minds, it was all about publicity, ego and advancing an agenda, not about science. Of course, a bombshell in science hardly stirs the air in the public arena; most of the impact is felt in scientific conferences or specialist journals. In fact, it's really all about what gets published, what doesn't and why; that is where the conflict gets serious.

The backstory was covered thoroughly by Gary Taubes in an infamous article published in *Discover* magazine. It had an impact for sure: some of Prusiner's colleagues complained that it amounted to character assassination, and Prusiner himself virtually gave up doing media interviews.[1] Taubes himself admitted later that he could have been a little less sardonic. But he had no difficulty finding scientists who were willing to comment, many of whom had worked in Prusiner's lab.[2]

One, Frank Masiarz, was to be a co-author of the controversial 1982 paper but eventually opted out. Masiarz felt that naming something implied that you had identified it, that you knew exactly what it was, and said that simply wasn't the case. Other scientists who were working with Prusiner at the time testified that "prion" became the mantra of the lab, and they suspected that coming up with a name and suggesting there could be connections to other neurological diseases was as much a ploy for funding as anything else.[3] But most of the resentment came from scientists who for years had been studying scrapie, who worked in institutions where it had been studied for decades, especially in the United Kingdom, and who thought Prusiner was a headline-hogging upstart.

The widespread negativity, especially from researchers in other labs who had been studying these disease agents for decades, simmered just below the surface of their otherwise science-y writing. An anonymous commentary in the medical journal *The Lancet* (which turned out to have been written by Alan Dickinson, one of the inventors of the virino hypothesis) chided Prusiner indirectly, in an academic style, by invoking the memory of Nobel Prize winner Wendell Meredith Stanley. In the summer of 1935, at a time when the chemical makeup of viruses was still a mystery, Stanley created a sensation by seeming to show that viruses could be composed of protein and nothing else. But he had been too hasty. He had ignored the relatively small amount of RNA (the cousin of DNA), about 6 percent, in his preparations of the particular virus he was working on, and fixated instead on the 94 percent that was protein. Dickinson, in referring to Prusiner, remarked that this sort of "half-truth" should be remembered in the current context.[4]

Nor was Dickinson as impressed with the new term for these agents as its inventor was, referring to it only once, dismissively, by setting it off in quotation marks: "prion." You can feel the chill in the academic air. Dickinson argued that the various different versions or strains of scrapie he worked with could be explained only by the presence of genes. Prusiner hit right back a few months later, dismissing the importance of strains and making sure to describe them as "strains." Quotation marks at twenty paces!

Carleton Gajdusek himself characterized those who were arguing for a new kind of infectious particle made solely of protein as "the 'romantics,' who have prematurely contended, on insufficient evidence, that it is a totally new form of replicating microbe, a pure protein."[5] More than one critic wondered what the fuss was all about. After all, Prusiner hadn't really added anything to the debate because he hadn't totally done away with the idea that there might be DNA or RNA in there somewhere. If you say that protein must be important, you're not hurting anyone's feelings, because decades earlier it had become clear that viruses themselves were some sort of gene kit inside a protein coat, that "bad news wrapped in a protein." On the other hand, if you make your case that it is *only* a protein, not only have you stepped on people's toes, you've also put pressure on the entire structure of molecular biology. Prusiner did neither because he still allowed for the possibility that genes played a role. But perceptive critics saw that it was pretty clear which way he wanted to go.[6]

But it wasn't just that Prusiner was talking heresy, if not actually committing it; it was the way he went about it. His critics were bothered by several things: his habit of ignoring data that didn't nicely fit with what he was proposing, his clever use of language to make the case where scientific data couldn't and a troubling tendency to claim discoveries for himself that seem to have been made by others.

For instance, before Prusiner even published his landmark 1982 paper, a researcher named Patricia Merz, an extremely skilled microscopist, had seen peculiar fibrils in preparations made from scrapie-infected brains, fibrils that she never saw in control preparations from undiseased brains. As the controversy over the prion claim was heating up, Merz published a paper, with colleagues including Carleton Gajdusek, showing that the same fibrils could be seen in mouse brains infected with CJD and kuru, but not in a large series of diseases that might seem superficially similar, like Alzheimer's and Parkinson's. The question was, were these microscopic fibrils the cause or the result of these diseases? She then showed that similar diseases causing equivalent tissue damage lacked them, but fibrils were plentiful everywhere in scrapie infections, even in tissues like the spleen that otherwise seem relatively unaffected. The consistency of her sightings and the careful elimination of other possibilities persuaded her and her collaborators to argue that what she was seeing was the actual agent of scrapie or kuru or CJD.

By now, Stan Prusiner's lab was at full throttle, and scientists there too were closing in on some unusual rodlike structures that they had found in scrapie-infected tissues. But rather than admitting that these were—or even allowing that these *might* be—Patricia Merz's fibrils, Prusiner argued that they were different, and that his were the true infectious agents. He has maintained that stance, even while most scientists believe that Merz made the original authentic discovery. There are other, even more inflammatory stories, most of which have been highlighted in Gary Taubes's *Discover* article or Richard Rhodes's *Deadly Feasts*. But clearly, even if only some of them are accurate, Stan Prusiner was one determined—some would say ruthless—scientist.[7]

It was a campaign for the minds of researchers. You'd think that the quality of the data was paramount and that a scientist's influence would be dependent on it and it alone. But science isn't so narrowly objective; Prusiner's fondness for his word "prion" reflects his appreciation of the power of language to sell scientific ideas, and his success in doing that has been noted by other scholars.

Prusiner structured that famous 1982 paper so that the protein-only suggestion, even though there were no conclusive data going for it, and even though Prusiner himself admitted it might not be right, was nonetheless presented as the most *exciting* possibility.[8] The idea that the scrapie agent might be a protein and nothing else was—his word—"heretical" (whoa, that sends a chill up your spine); if true, the consequences were "significant" (who could resist?); and "a knowledge of the molecular structure of prions may help identify the etiologies [causes] of some chronic degenerative diseases in humans" (well, now my jaw just dropped!). Prusiner was in effect saying, There's a lot of confusion out there, we don't really know what we're dealing with, but the protein-only train is leaving the station and you should be onboard.

Prusiner is a magician, diverting the reader away from the scant evidence for his claims and holding out instead the promise of those claims, influencing the direction of others' research as a result and making the likelihood of finding genes associated with these agents gradually fade away.

Yet, not everyone was persuaded. Some scientists who have disagreed with Prusiner from the beginning argue that the pre-eminence of prions is a triumph not of good science but of the power of language. The most prominent of those is Dr. Laura Manuelidis of Yale University. In a review of Prusiner's book *Prion Biology and Diseases* in 2000, Manuelidis wrote: "There are indeed many extraordinary aspects of this new biology of prions, not the least of which is their possible non-existence." And later, "*Prion Biology and Diseases* is stimulating in ways probably not anticipated by its authors. It leads one to reexamine the objectivity of science and whether it is a myth vanished. It underscores the stunning force of the declarative sentence and, although I hate to admit it, the peculiarly American sport of betting on popular momentum."[9] Manuelidis stands out from the rest of Prusiner skeptics in that she is still waging that war, even today, and she is one of the few scientists who claim to have concrete evidence that prions are not what they are supposed to be. We will meet her again later.

What about the man at the center of this storm? Prusiner gave as good as he got, both in the scientific literature and in person. But he offered few insights into his own feelings during the height of the controversy. However, here, in this excerpt from his autobiography on the Nobel Prize website (Prusiner won in 1997), things are different:

> Publication of this manuscript, in which I introduced the term "prion," set off a firestorm. Virologists were generally incredulous and some investigators working on scrapie and CJD were irate. The term "prion" derived from protein and infectious provided a challenge to find the nucleic acid of the putative "scrapie virus." Should such a nucleic acid be found, then the word "prion" would disappear! Despite the strong convictions of many, no nucleic acid was found; in fact, it is probably fair to state that Detlev Riesner and I looked more vigorously for the nucleic acid than anyone else. While it is quite reasonable for scientists to be skeptical of new ideas that do not fit within the accepted realm of scientific knowledge, the best science often emerges from situations where results carefully obtained do not fit within the accepted paradigms. At times the press became involved since the media provided the naysayers with a means to vent their frustration at not being able to find the cherished nucleic acid that they were so sure must exist. Since the press was usually unable to understand the scientific arguments and they are usually keen to write about any controversy, the personal attacks of the naysayers at times became very vicious.[10]

Who knows at what point skeptics become "naysayers," their search for genes leads to "frustration," those elusive genes become "the cherished nucleic acid that they were so sure must exist." Remember that Prusiner himself was very, very careful, at least at first, to allow for the possibility of that "cherished" genetic material. Here's how he described prions in a 1984 *Scientific American* article: "Even if some DNA or RNA is ultimately found in the protein . . ."[11]

Fast-forward a few years and you'd never suspect he had had any doubts that prions were proteins and nothing else: "Fifteen years ago I evoked a good deal of skepticism when I proposed that the infectious agents causing certain degenerative disorders of the central nervous system in animals and, more rarely, in humans might consist of protein and nothing else. At the time, the notion was heretical."[12]

This was the setting for the birth of the science of prions. The scene has changed dramatically since Stan Prusiner named them in 1982. Most of the scientific community has come onside and there's a tremendous amount of very cool science being done, Prusiner has his Nobel (making the Nobel committee either prescient or presumptuous) and there's a growing conviction that this represents a whole new kind of science, a science whose full impact has yet to be felt. So it's time to move on from the controversy and bitterness and see just where that new science is taking us. There still are skeptics; there remains a slim possibility that prions are not what the majority of scientists think they are, but let's set those doubts aside for the moment.

13

A Portrait of the Prion
And the Experiments That Point to Their Role in the Human Brain

Kurt Vonnegut's *Cat's Cradle* contained more than a telling reference to the importance of proteins. In the novel, Vonnegut invented a mysterious substance called ice-nine. It was dangerous stuff.

> "But suppose, young man, that one marine had with him a tiny capsule containing a seed of ice-nine, a new way for the atoms of water to stack and lock, to freeze. If that Marine threw that seed into the nearest puddle?"
> "The puddle would freeze?" I guessed.
> "And all the muck around the puddle?"
> "It would freeze?"
> "And all the puddles in the frozen muck?"
> "They would freeze?"
> "And the pools and the streams in the frozen muck?"
> "They would freeze?"
> "You bet they would!" he cried.

Ice-nine crystallized in a different way from the water we all know and stayed solid until it was heated to 114.4 degrees Fahrenheit (45.78 degrees Celsius). A few crystals of ice-nine would freeze ordinary water instantly if they touched it—their peculiar crystal structure locking in all the water molecules with which it came in contact.

For a brief time in the 1960s, it looked like ice-nine had a real-life counterpart: Soviet scientists announced they had discovered an anomalous form of water—it was syrupy, with a higher boiling point and a lower melting point than ordinary water. This strange water, which was eventually named "polywater," was thought to owe its unusual properties to the fact that it was a polymer, a short chain of water molecules. In the distinguished pages of the journal *Nature*, F. J. Donahoe wrote:

> I need not spell out in detail the consequences if the polymer phase can grow at the expense of normal water under any conditions found in the environment. Polywater may or may not be the secret of Venus' missing water. The polymerization of Earth's water would turn her into a reasonable facsimile of Venus.
>
> Only the existence of natural (ambient) mechanisms which depolymerize the material would prove its safety. Until such mechanisms are known to exist, I regard this polymer as the most dangerous material on earth.[1]

Fortunately for the earth, polywater turned out to be a mistake, a result of tiny amounts of contamination that changed the properties of quite ordinary water into something that looked spectacular. So, in the end, polywater, like ice-nine, was fiction too.

But if scientists are correct, the way prions reproduce relies on a similar sort of contagious chemistry. It has been a struggle to win a skeptical scientific community over to the idea that something this outrageous is actually a fundamental part of biology, and there are still unanswered questions about prions. Big ones, like how exactly do they cause disease, how precisely do they multiply and, at least in the minds of some, how do we even know they exist?

There remain some scientists who think alternative scenarios from decades ago still fit the data nearly as well as do prions. But the number of those who continue to argue for a virino, or even some reasonable facsimile of a virus, has gradually dwindled. On the other side, there is a solid consensus, one that has grown significantly since Prusiner first "invented" them, that prions are the real deal and that they represent a totally new frontier in medicine and biology. So, while acknowledging that there is still a handful of scientists who don't believe in them, here is a portrait of the prion.

They are indeed proteins of an average size, about 250 amino acids long, with a few other little clusters of atoms here and there. Nothing remarkable in any of that. Even though early on it seemed that prions had no DNA in them—no genetic code *of their own* that would make them infectious—they were still proteins, and so somewhere there *had* to be a DNA code that would direct their synthesis. Once again we have Stanley Prusiner to thank for finding that code. Once that skeletal 250-

amino-acid sequence had been determined, Prusiner enlisted the help of some high-powered experts in genetic engineering to search around inside scrapie-infected cells looking for any DNA or RNA that might represent the code for that protein.

The technology they used actually followed Francis Crick's central dogma but in reverse: tracing the origin of a protein by changing the direction of the arrows: protein → RNA → DNA. Not making the DNA from the protein, but using the protein to make a probe that is then used to find the DNA.

There were two stunning discoveries made in the course of these experiments. For one thing, there was indeed DNA in diseased cells that coded for the prion protein. That was a little troubling for the protein-only idea. First, prions are infectious agents, which usually implies they invade cells from the outside. Why would there be a DNA code for them waiting inside the cell? Where did it come from? Proponents of the idea that prions were in fact peculiar viruses must have taken heart upon hearing this, because that's exactly what you'd find in a virus-infected cell: proteins (pieces of the virus) accompanied by bits of DNA (the virus genes).

But that discovery wasn't the only wrench in the works: those same researchers were dismayed to discover that there was also the same DNA in *healthy* cells. Now, that made no sense at all! How could the code for a disease-causing agent be resident in body tissues when they're healthy? Prusiner himself worried about that and thought he and his colleagues might have made a "terrible mistake."[2]

Put yourself in his place: you're on the track of a sensational new kind of infection, you've named it (incurring the wrath of other scientists), you practically *own* it, and now you've found that the blueprint for making it is right there in completely normal cells exhibiting no signs of disease. Prusiner wondered if the way out of this paradox was that there were actually two forms of the protein made by this gene, one causing fatal disease, the other benign and normal.

This is radical: if there were two forms of this protein—one good, one bad—they had to have exactly the same amino acid sequence, since they had been found using exactly the same DNA code. Identical amino acid sequences should make the proteins themselves identical, and yet, if Prusiner was right, they came in two varieties. As he and his colleagues worked their way through experiment after experiment, they came up with some evidence for the two-protein idea: they were able to show that the disease-causing version of the protein was resistant to an enzyme that normally breaks down protein. In contrast, the benign version was, like most proteins, susceptible to that enzyme. That could only be the case if the difference between the two boiled down to some difference in their structure, and because it couldn't be the sequence of the amino acid chain, it had to be the way it folded up.

This strange discovery challenged the most basic biological thinking. Could it be that, unlike any other protein known to science, the unique amino acid sequence of the prion was somehow able to adopt two different final shapes? And could that happen with a momentum and ferocity that allowed the disease-causing twin to multiply and eventually cause disease and death? Moreover, if the harmless version of the prion protein was to be found in healthy cells (that's why it was defined as harmless), then what did it do?

In science, the trick is to frame ideas as questions that can be answered in the lab. Those questions were asked, and there were indeed answers.

The key question was, could the disease-causing version of the prion protein be folded in a different way that made it resistant to digestion by enzymes? Yes, it is *misfolded*. Even today, no one is sure exactly how different the misfolded version is from the normal version, because the misfolded protein aggregates with others of its kind so aggressively that the individual molecules are just too difficult to separate. That makes it impossible to apply the standard biochemical technologies, like X-rays, that could create detailed, atom-by-atom images of the structure. But the few shadowy glimpses of the misfolded prion protein that exist have revealed that where the normal version is a complex of twists and turns of coils, misfolding creates a molecule that has abandoned the coils in favor of back-and-forth turns creating sheets. These sheets account for the resistance to breakdown by enzymes.

This discovery was a shocking turn of events. From the time protein folding was figured out, it was generally accepted that each protein should have one ideal, stable shape, with the fewest possible stresses and strains built into the structure. Proteins do what they do by virtue of their shape, and so the shape each takes as it is made should be consistent.

Remember hemoglobin? The way in which it is folded shelters its precious iron atom from potential chemical disruption

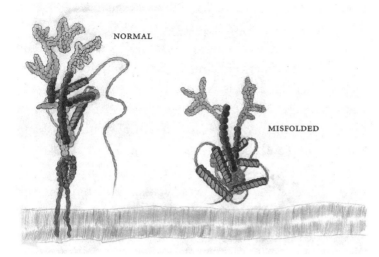

On the left, at very high magnification, is a single prion protein folded in the "normal" way, although there are probably many slight variations on this theme, none of them harmful. Even this one might shift slightly if we were to watch it for a while. On the right, the misfolded version. This isn't just a variation on a theme—this is a full remix, with disastrous results. Vast stretches along the molecule have changed shape and orientation.

but at the same time exposes it in just the right way so that it can exchange oxygen and carbon dioxide in exactly the appropriate circumstance. Hemoglobin is only one protein: the entire suite of chemical reactions in the body is managed by the proteins called enzymes, and each has what's called an active site, a pocket or slot or cavern where the molecules participating in that reaction come together. Most of the time, alterations to the shape of that active site would make the enzyme useless. In fact, shape reigns over amino acid sequences, and the only possible substitutions to that sequence are those that don't substantially disturb that shape.

The word "substantially" is key. "One protein, one shape, one function" is an attractively simple picture, but nothing in life, even at the cellular level, is ever simple. It's sometimes possible for a protein to adopt more than one "final" shape, and the fact that the prion protein could take two radically different shapes, and maintain them, proved that.

If the idea that an amino sequence could fold in two distinct ways wasn't crazy enough, it was even more stunning to realize how that apparently happens. In the extraordinary case of prions, the misfolded proteins somehow convert, or recruit, normal ones. A normal and an abnormal come together, touch (although no one knows exactly how) and two abnormals emerge. How that happens is also mysterious, but we can get a rough idea by thinking about the common mousetrap. It exists in two different forms (or shapes): sprung and unsprung. The transition from one to the other is rapid and definitive. Fill a room with them, as was once done in a Walt Disney program on nuclear fission, and make sure they're all just on the cusp of springing. Then toss a sprung (or misfolded) one in: it collides with a trap on the floor, the second trap springs (misfolds) and the process is on its way. Note that the process does not go in reverse.

This is an absurdly simple version of what actually happens, but just how absurd is unknown, because it's a mystery how the rogue versions of prion proteins misfold. As I said, they clump together so aggressively that the individual proteins just can't be distinguished. But it does look like these clusters of misfolded prions create the rod-shaped fibrils that Patricia Merz identified back in the early 1980s.

There are two significant things about these fibrils. First, if they continue to grow in length and stick together, they apparently become the dark blobs that Igor Klatzo remarked on in kuru brains, and that are seen, although in slightly different versions, in other prion diseases. On the other hand, it doesn't seem that these foreign objects, these Twizzlers of prions, are infectious. If they're broken up into smaller pieces, however, maybe a dozen or so rogue proteins in size, they are. To review this hierarchy: the fundamental piece is a single misfolded protein; thousands of those stack together like Lego to form fibrils and thousands upon thousands of fibrils are jumbled together to form plaques. The only infectious pieces are broken-up pieces of fibrils, mere handfuls of misfolded proteins.

What might these small clusters of misfolded prions be doing? Think of ice-nine. One of these clusters could become a scaffold onto which newly recruited prions are pressed into place—misfolding as they attach—each one a piece of the growing fiber. If this were a jigsaw puzzle, when you tried to press the wrong piece into an available space, it just wouldn't go. But in this case, the wrong piece (a normal prion protein) will conform by misfolding to the space into which it's being pressed. And where might this happen? Prion proteins—the normal ones—are embedded in the surface of the cells they inhabit (especially neurons, or brain cells). Experiments have shown that this attachment is necessary for infection and destruction: free-floating prion proteins don't get recruited by misfolded ones. So it seems that the pathological process hap-

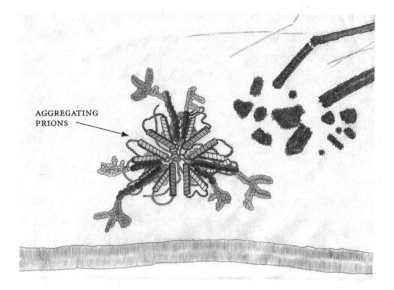

Misfolded proteins begin to aggregate and eventually dozens, even hundreds, cluster together to form fibrils that, as they fall apart into fragments, recruit more normal versions. This diagram is highly speculative. We just don't know the details yet.

pens somehow at the cell surface. Imagine the normal, healthy situation: a neuron, a long, spidery brain cell with multiple extensions, its surface studded with properly folded prion proteins, especially around the synapse, the area where the neuron communicates with its neighbors. Enter the misfolded prions. Some of them are solo; many are clustered with others of their kind. They are drifting with the molecular winds; their counterparts are fixed in the cell membrane. Occasionally, some sort of contact is made by a cluster with a normal prion, and each time that happens, there's a conversion event.

Unfortunately—very unfortunately—it's a one-way process. Normals never convert the misfolded. But misfolded prions definitely convert normals. They *are* ice-nine.

The intimate details of this transformation are unknown. Do the two come into direct physical contact and then part, with the normal prion protein bearing the misfolded imprint of the exchange? Is the new rogue protein always joined to a growing chain of others? Do other molecules intervene in the process, acting as intermediaries to make it possible? Remember, the normal folded shape is presumed to be energetically stable, and there should be some sort of energy hill to climb if it's to switch, and it might not be able to do that by itself. Stanley Prusiner and his colleagues once suggested there was a third party, a chaperone that they called "protein X." It has never been identified.

Of course, all these questions bear on the most important point of all: What is it about this process that causes disease and inevitable death? This is where two different investigations become entangled. One is the search for the cause of damage; the other is the effort to understand what normal prion proteins do and why they are so populous in the cells in our brains. It's almost impossible to pursue one avenue without involving the other. For instance, maybe the destruction caused by these diseases is the result of the accumulation of misfolded prions, big globs of junk—plaques—that are seen in the brains of those killed by disease. You'd think that would be extremely disruptive in the highly complex and ordered setting of living cells. On the other hand, maybe it's simply the gradual, but unstoppable, depletion of normal prion proteins in the brain.

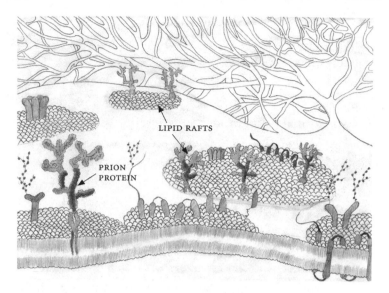

Especially prominent on the surface, if this cell were a neuron, would be these: normal run-of-the-mill prion proteins embedded in "lipid rafts" on the surface of the neuron.

To address that question you'd have to know what their job was in the first place.

A beautiful experiment performed in the early 1990s shed light on both of these issues. Researchers bred genetically engineered mice that lacked normal prion proteins. (Lab mice can easily be bred lacking specific genes: they're called knockout mice.) The resulting mice, lacking the gene for making prion proteins, were immune to prion diseases. They could be pumped full of scrapie prions and remain completely normal and unaffected. It fit perfectly with theory: even enormous doses of invading prions aren't sufficient—by themselves—to cause disease. They must recruit many multiples of their own numbers, and they can't do that if there aren't any normal prion proteins there.[3] But the experiment also showed that accumulations of misfolded proteins weren't, by themselves, toxic.

The point was underlined by experiments where grafts of brain tissue with the normal complement of prion proteins were introduced into the brains of mice engineered to lack them. Then misfolded prions were injected into those hybrid brains. The grafted tissue, as expected, became diseased and released large numbers of misfolded prions, which migrated throughout the brain, even into the areas where the brain cells lacked prion proteins. Those cells still resisted infection.

But these experiments also revealed something else that was particularly hard to swallow. The mice engineered to have no prion proteins in their brains (or anywhere) didn't seem to miss them. This flew in the face of the knowledge that these proteins are highly evolutionarily conserved; that is, that they exist in roughly similar forms in a virtual zoo of species.[4] Living things don't waste energy making proteins that aren't useful, but as far as anyone could tell, the mice developed normally; very little seemed to be wrong with them other than some subtle memory deficits. When the experiments were extended to cows, the same strange result was obtained: at least for the first three years of life (as long as they were allowed to live), the cows seemed completely normal, at least behaviorally. They had lost some specific kinds of cells in a part of the brain called the cerebellum, but you wouldn't have known without autopsies of their brains. Superficially, these were normal cows. The researchers have now used bull semen to establish an entire herd of cows with no prion proteins.

These findings, showing that a complete lack of prions seemed to have minimal effects, were so counterintuitive that they precipitated a rush of experiments, establishing a new, different, but just as mystifying picture: whereas before, no role for prion proteins could be assigned because their removal seemed to have no effect, now it's impossible to tell what they're responsible for because there are so *many* effects. Bearing in mind that these experiments are hugely challenging technically, and so the results in many cases have to be considered tentative, there is now evidence that the absence of prion proteins (in mice, at any rate) alters circadian rhythms, the sense of smell and the transmission of nerve impulses, and causes an increased number of seizures and deterioration of peripheral nerves (the nerves outside the brain).

Experiments also suggest that they play a role in regulating cell suicide, called apoptosis, and that they might prevent oxidative damage. Because normal prion proteins are embedded in cell membranes, they likely play a role in receiving or translating the molecular signals that move among cells. Prion proteins are even implicated in the normal function of the immune system. So are we further ahead? Well, yes, in that there had to be a role for them, and finding none at first didn't make sense—but right now, there is an embarrassment of riches.

There is one possible role as yet unmentioned that may be the most intriguing and, in the long run, the most important.

And this is that the normal prion protein might play an important role in creating long-term memory. The evidence of this is diverse, maybe too diverse to be easily connected, but there are some strong hints. For instance, in one experiment, mice were trained in a maze that resembled a clock face, except that instead of numbers the periphery was lined with twenty small holes, only one of which led to a tunnel that allowed escape. After becoming accustomed to the maze, each mouse was released from a chamber in the center of the clock and rated on how fast it could locate the tunnel. On subsequent tests, the mouse was evaluated on how much it remembered: Did it just search the clock face randomly; did it search systematically, moving along the periphery from one circle to the next; or did it make a beeline for the tunnel without delay? Mice genetically engineered to lack prion proteins didn't improve with time, meaning they weren't remembering their previous trials. But one particular group, whose prion deficit was reversed partway through the experiment by a time-dependent gene switch, performed poorly up to the time of the switch, but then started to perform better, showing that once their prion proteins were functional, their memories improved.[5]

At the other end of the experimental spectrum, genetic studies in humans have shown that certain kinds of mutations in prion protein genes affect long-term memory. Swiss students were shown a series of nouns, one per second, then asked to remember them five minutes later, and again a day later. At the same time, they were screened genetically. Alterations in their prion protein genes (at one single amino acid site in the entire protein) correlated with as much as a 17 percent difference in long-term recall of the words. These genetic differences had no effect on short-term memory.[6]

There is a hint of something coming together here: prion proteins exist in profusion at synapses, the places where brain cells meet, and where it's assumed changes would have to occur to establish long-term memories. Because of their notorious ability to change shape and remain changed permanently, they could be agents of such change. Indeed, there's evidence that a large sea slug of the genus *Aplysia* apparently uses prionlike molecules to establish such memory. Add that to the mouse and human experiments and you can see glimpses of a pattern. But it is still too early to be able to pin down exactly what the normal prion protein does in our brains. And without that knowledge it remains difficult to establish just how the rogue versions of the protein cause disease. In the long-term-memory scenario, being able to alter protein folding is a good thing. In prion disease, it goes wrong. Just how dramatically wrong is the subject of the next chapter.

14

Mad Cow Disease
The Uncertain Ground Where Politics and Science Intersect

In a way, every twist in the story of prions before the 1980s was mere prologue. That decade not only landed Stanley Prusiner on center stage but also brought us BSE, better known as mad cow disease: "b" for bovine, "s" for spongiform (because a diseased brain is riddled with microscopic holes) and "e" for encephalopathy, or brain disease. One of the most remarkable things about the epidemic of mad cow disease was the contrast between the speed with which the disease spread and the sluggish human response. For "human" read "political": mad cow was a classic example of how politicians will twist, conceal and bury scientific knowledge until the downside of silence and concealment threatens their political life more than the upside. But only then.

We will likely never know when the first English cow contracted mad cow disease. A definitive description of the disease wasn't published until 1987, but it's pretty clear that cows were dying of the disease years earlier. They weren't diagnosed with BSE because their unique suite of symptoms—clumsiness, aggressive behavior, confusion—had never before been recorded and were simply ascribed to some sort of unusual toxin or infection. But when you put all these early cases together, you get a stunning picture of a disease that even in its early stages promised to be catastrophic.

In December 1984, veterinarian David Bee examined Cow 133 at Pitsham Farm in Sussex, owned by Peter Stent. When Bee was called in, the cow had lost weight and had an arched back. She went on to develop a head tremor and lose her coordination, and in February 1985 she died. By the end of April, five more cows at the farm had died, and veterinarians were using the term "Stent Farm Syndrome," but they had no clue what the cause of death was. Bee thought there was some sort of fungal toxin contaminating the feed at the farm, but when the remains of those cows were examined, the results were inconclusive.

In Wiltshire, hundreds of miles away, between October 1983 and May 1985, a vet saw at least five cows on Paul Lysley's farm. Same symptoms: lack of coordination, weight loss, hyperexcitability or aggression. It wasn't until years later, when the vet sent a cow from the same farm to the Central Veterinary Laboratory and it was found to have had BSE, that it became clear these earlier cases were likely BSE too.

In April 1985, a cow was seen to be acting weirdly on the Plurenden Farm in Kent. One cow soon became five. Their brain tissue was examined in detail, revealing that spongy transformation and destruction of neurons, and the examiner commented that the damage resembled that caused by scrapie, only with subtle differences. That was the first, but not the only, suggestion of a link between scrapie and BSE. When two additional cows at Plurenden were found with similar symptoms, one was shipped to the Central Veterinary Laboratory to be euthanized and necropsied. A full report of the damage to Cow 142's nervous system was made in late September 1985 by pathologist Carol Richardson. In her written report, the first line of diagnosis reads, "moderate spongiform encephalopathy—acute."[1] Years later, and much more dramatically, Richardson testified in writing to the BSE Inquiry in the United Kingdom that, as far as she was concerned, this was scrapie in a cow. She also testified that she had discussed the similarity with a colleague and that investigators had been aware there were other such cases. Given that there was no official acknowledgment for many months afterward that mad cow even existed, this had the really bad odor of a cover-up.

But the official BSE Inquiry failed to find evidence that these intriguing claims had any real basis. If they had, there would be hard questions to be asked as to why the official notification of a new disease in cattle didn't come until 1987, two years later. The inquiry decided that Ms. Richardson's memory was faulty and that in reality no one had agreed on and acknowledged the scrapie-like character of these early cases.

But even if Carol Richardson hadn't provided the smoking gun, there was plenty of evidence revealed in the inquiry to show that "cover-up" was a perfectly legitimate word to use in describing what was happening in that first year and a half of BSE. On the one side were veterinary scientists who were becoming aware that there was a new disease in cows, one that the public knew nothing about. And these weren't just a few concerned James Herriot–like gentlemen sharing their obser-

vations over a pint in the local pub. On the other side were the
government people, scientists working in agencies like the Veterinary Research Laboratory, who were coming to the same realization but were vividly aware of the effect an announcement
of a new, scrapie-like disease in British cattle could have on
the export beef trade. There were already countries that were
unhappy that scrapie was doing nicely in Great Britain, even
though there were no data, then or now, to show that eating
a scrapie-ridden sheep was a danger to your health. And now
cattle with their own version of scrapie as well? That would be
really bad for the Union Jack.

As the number of farms with BSE on them began to grow,
alarmingly quickly, the number of people with inside knowledge
grew. But they had very different ideas of what to do. The disease
was volcanic, the government response glacial. For instance,
veterinarian Roger Hancock, intending to publish a 170-word
article in the *Veterinary Record* about one of the first instances
of BSE in England, had written to Bernard Williams, head of
the Veterinary Investigation Service, with a description of a
cow's brain with the all-too-familiar signs of a spongy-looking
brain and holes in neurons: "A diagnosis of spongiform encephalopathy of unknown aetiology was made. The lesions
were similar to those seen in sheep with scrapie."[2]

Uh-oh! Scrapie. Having first agreed to the publishing of
the article, Dr. Williams changed his mind and in a letter explained to Hancock that "because of possible effects on exports
and the political implications it has been decided that, at this
stage, no account should be published."[3]

Politics and exports. Funnily enough, when Williams testified to the BSE Inquiry years later, he recalled that he didn't want to publish Hancock's paper because he thought it was uninformative and incomplete.

Then it was Gerald Wells's turn. The head of the Consultant Pathology Unit, he proposed a two-part paper for the *Veterinary Record:* the first half would be a description of the symptoms in cows and the abnormalities seen in microscopic images of their brains, the second half a comparison between this, scrapie and other related spongiform diseases. Wells was told that only the first half of the article was acceptable. He— and others—objected to this corruption of science by politics but was eventually persuaded that there were still some weaknesses in the link from the cow disease to scrapie, and the paper was postponed.

That word "scrapie" had become an obsession. In the summer of 1986, a nyala, an African antelope, had died at Marwell Zoological Park after behaving weirdly for a couple of weeks. Dr. Martin Jeffrey diagnosed a spongiform encephalopathy, remarking that the similarity to scrapie was striking and, unaware that Carol Richardson had said the same thing about a cow, claiming it was likely the first such disease ever observed in a bovine. It was good science; one expert who read the report thought the pathology Jeffrey had recorded was dramatic. But Dr. Jeffrey wanted to call his article "Scrapie-like Disorder in a Nyala." He pushed the wrong button with that and was informed that approval to publish was unlikely if comparisons were made with scrapie. The failure to publish would be negligent, he felt, but nonetheless he was forced to wait. Eventually, when the paper was published a full two years later, in 1988, it referred coyly to "a provisional diagnosis of a scrapie-like disease."[4]

But even among the gatekeepers there was growing discomfort about the choke hold on publications describing the new disease. Were they becoming worried about the inappropriate censoring of science? Actually not. They were concerned that someone else would beat them to the punch and get credit for the identification of BSE. "Here is a golden opportunity," one wrote, "for the Veterinary Investigation Service and Central Veterinary Laboratory to demonstrate to the world that we are performing the function of identifying and investigating new conditions in farm livestock. It would be a great pity if we did not receive the recognition. It would be worse if another group appeared to be successfully carrying out our function."[5]

Even though the evidence was mounting that this cattle disease was a lot like scrapie, and most scientists agreed that it had to be made public, there was still resistance. John Suich, of the Animal Health Division of the then Ministry of Agriculture, Fisheries and Food, wrote British embassies around the world at the end of October 1987: "It would be particularly misleading if it were to be described as 'scrapie in cattle.' Scrapie is a disease of sheep. . . . A point to emphasise, if you are pressed on numbers of cases, is that while it may be *suspected* in over 100 herds and distributed over a wide area, it has been *confirmed* in only 25 animals, out of a total UK cattle population of just over 12.5 million."[6] In fact, between May and December 1987, the number of suspected cases rose from 13 to 370, and confirmed cases rose from 6 to 132. In the same period, the number of herds involved rose from 4 to 237.

It was all a sordid mix of politics, economics, self-interest, the desire to be first, the desire to inform scientists in other countries and an honest feeling that science shouldn't be published prematurely. But politics had the whip hand. And while even the handful of examples above would boil the blood of anyone fond of freedom of information, the political machinations were just starting. Even so, one intelligent decision had been taken back in June 1987: epidemiologist John Wilesmith was assigned to figure out what was going on. He was given the details as they were known then: six confirmed cases of the disease on four farms, a probability that there had been cases as early as 1985 and that it was like scrapie. That the disease had appeared in herds in widely different areas of the country would ordinarily be taken to mean there wasn't a link among them, but the disease seemed to have started more or less simultaneously in all these places, which would suggest there might indeed be some sort of link.

It wasn't much to go on. Wilesmith was in roughly the same position as Carleton Gajdusek and others had been in New Guinea three decades before: he knew he had a serious disease on his hands, but the cause was almost completely mysterious. Wilesmith did have one advantage over the earlier investigators: he knew that BSE was like scrapie. The suspicion that the sheep agent might have jumped to cattle led him and his team to begin checking out possible mechanisms by which that could happen, which in turn led them to investigate cattle feed.[7] At the time, dairy cows had their protein intake boosted by feeding them something called meat and bonemeal, derived from the industrial processing of the remains of a variety of dead animals. It was a way of boosting their milk production and a huge business in Britain. More sheep remains than ever were entering the cattle food chain, and from what Wilesmith could tell, the scrapie agent would survive the processing used to prepare the stuff. (This is where all those attempts in the 1960s to kill off the scrapie agent in the lab began to yield an unexpected payoff.) All the affected cows, regardless of their location, had been fed meat and bonemeal. All were dairy cows, significant because beef cattle weren't generally given those supplements. Also, meat and bonemeal had been part of the diet of Jeffrey's famous nyala and another antelope, a gemsbok, both of which had died with brain lesions reminiscent of scrapie.

But suspecting meat and bonemeal as a conveyance for scrapie didn't explain why the BSE epidemic started so abruptly and spread so widely. Meat and bonemeal had been given to cattle for years prior to the outbreak. Had anything significant changed to precipitate this disaster in the early 1980s, the date which Wilesmith calculated must have been the time when the seeds of BSE were sown? (Remember, scrapie, like all these diseases, had an extended incubation period, so cases appearing for the first time in the mid-1980s must have been the result of infection years earlier.)

Wilesmith thought it unlikely that he was looking at a new mutant strain of scrapie predisposed to infect cattle, because the new disease had appeared out of thin air almost simultaneously in farms hundreds of miles apart. He was convinced that something had happened to ramp up cows' exposure to the scrapie agent, and he identified several factors that he thought might have done that, including a sharp increase in the sheep population, a concurrent increase in the number of cases of scrapie, the inclusion of more sheep heads in the mix of stuff that was turned into meat and bonemeal (the brain of a sheep that died of scrapie would be full of the scrapie agent) and a decrease in the use of solvents to process the stuff (to separate fat from protein), which might have allowed the scrapie agent to survive where it didn't before.

Wilesmith came pretty close to hitting the nail on the head, identifying meat and bonemeal as the vector responsible for the outbreak of BSE, but his idea that changes in the processing of meat and bonemeal led to the nearly simultaneous infection of many cows by a scrapie agent that had always been there (although likely at lower levels) didn't strike the BSE Inquiry years later as having the ring of truth. First, further investigation of carcass processing made it clear that even the earlier, supposedly stricter procedures, wouldn't have inactivated the scrapie agent. The inquiry also had a hard time believing in the sheer simultaneity of it all and found it unlikely that the plain ordinary scrapie prion had suddenly been unleashed. Instead, the inquiry committee suggested, BSE was caused by a new prion.

Now, admittedly, given that they were examining the state of things in the late 1990s, those working on the inquiry had information at hand that Wilesmith didn't. Crucial information. For one thing, the BSE prion behaved very differently from the scrapie prion. There seemed to be only one strain— one version—of BSE, but there were more than a dozen strains of scrapie. It seemed unbelievable that only one of those would have succeeded in passing from sheep, through meat and bonemeal, to cows in all the farms where it happened. Second, that singular BSE prion was different from scrapie in other ways. For instance, it didn't have the same pattern of success in infecting lab animals. Hamsters were easily infected with scrapie but were immune to BSE. Mink fed BSE came down with a mink version of the disease but resisted scrapie in their food. Of course, there were the exotic animals too, including the nyala and the gemsbok, but also a host of others, both her-bivores and even carnivores like tigers. All became infected after BSE was on the loose; all could presumably have been infected with scrapie long before but weren't.

It wasn't just the range of animals that were susceptible to BSE; their clinical picture was not the same either. The most heavily infested organs were different, as were the areas of the brain. The BSE Inquiry decided that the evidence pointed not to a jump of the normal scrapie prion from sheep to cows through meat and bonemeal but to either an altered scrapie or a novel cow prion. And finally, although the original outbreaks seemed to have been virtually simultaneous across the south of England, better and more detailed analysis of the data sug-gested that there could have been a single origin of BSE—a single infected cow, somewhere in the southwest, in Devon or Somerset. In other words, one cow that died, made its way into meat and bonemeal, infected others and kick-started the epidemic.

There were two related and very heavy implications of this change in thinking about the origin of BSE. First, if it wasn't the scrapie prion that had somehow snuck through the processing of meat and bonemeal in the early 1980s but a prion that had arisen somehow spontaneously in cattle, that could have happened much earlier—a truly unsettling thought. Think of it: one animal comes down with BSE in the 1970s, the disease is, of course, unrecognized as anything new (cattle succumb to superficially similar diseases all the time) and the cow gets sent to the rendering plant and is processed into meat and bonemeal.[8] (In fact, she might well have been slaughtered before she even showed symptoms, given that there is a lengthy incubation period.) Once she's been rendered, she's fed to other cows, her prions infect several of them and, after an incubation period of several years, they too fall ill, die and are recycled, followed by another delay for incubation, then more diseased cows. In this scenario, the cows of Pitsham Farm might represent not the first cases but a second or even third round.

Why does this matter? This is the second implication. Projections of the ultimate number of diseased cows might have been much higher if it had been thought that the diseases had been recycling for years. Also, and ultimately more important, as long as BSE was thought to be the scrapie prion residing in cows, the threat to humans didn't seem significant. After all, scrapie-infected mutton and lamb had been eaten for centuries with no ill effects, which led to the assumption that there was a significant species barrier that the prion simply couldn't jump. That species barrier had been breached in the lab by inoculating scrapie into a variety of lab animals (including squirrel monkeys), but those were laboratory manipulations after all, and it was easy to be confident that this same scrapie prion masquerading as BSE would have the same limited ability to jump from one species to another, especially to humans. But as the evidence mounted that this prion was different, an unsettling question emerged: If it could infect cows and many other animals, some of them by oral transmission, were humans completely safe?

This is where the two implications merge. If the BSE prion had been in circulation for longer than originally thought, unnoticed in cattle, and the prions in those cattle had heavily infested the brains, spinal cord and other organs of those cows, and the cows were sent to the slaughterhouse before any symptoms were apparent, then BSE prions had been entering the human food chain for a long time, much longer than originally thought. This unhappy thought turned out to have more truth to it than anyone realized.

15

Mad Cow in Humans
No Barrier After All

The eventual decline of BSE starting in the mid-1990s would have been much more satisfying if the numbers by then weren't so shocking. In the end, something like 180,000 British cows died of BSE, more than four *million* were slaughtered preemptively and somewhere between one and three million had been infected. First calculations suggested that no more than a gram of infected meat and bonemeal (famously described as two peppercorns' worth) was enough to infect a cow, but that number was later revised downward to an astounding one *milligram*. Normally, calves were fed ninety grams a day, and one infected cow, once rendered into feed, could infect ten others. If you're wanting a disease to spread quickly, that is the math you need. It was a runaway epidemic.

The first important step to slow it down was to prohibit the feeding of ruminants to ruminants. No more cow cannibalism. That ban took effect in 1988. But this didn't close the wound: both carelessness and deliberate lack of compliance conspired to ensure that the ban was anything but airtight, and thousands of calves born after the ban became infected. Actually, more than forty thousand. In retrospect it became clear that most of those were on farms where meat and bonemeal were still being fed to pigs and chickens. This was allowed because neither species was apparently susceptible to BSE. But evidently preventing the spillover of meat and bonemeal from their feed to cattle feed was much more difficult than anyone had predicted.

Fortunately, the original ban had already been tightened even before the first born-after-the-ban cow came down with BSE. In 1989, in recognition of the fact that cows could be infectious before they showed symptoms, those parts of the cow where prions were most often found—primarily, parts of the nervous system—were excluded from the human food chain. Then, when the notorious death in 1990 of a pet cat named Max showed that BSE was not limited to cattle, meat and bonemeal were eliminated (or supposed to be) from any kind of animal feed, including pet food. With those measures in place it seemed that the BSE epidemic was, if not over, at least heading that way. But it had been a body blow to the British livestock industry, with export trade interrupted and many thousands of cows either dying of BSE or slaughtered and incinerated.

And there were still those estimated one, two or three million cows, infected but undetected, the majority of which had been slaughtered and fed to humans before exhibiting any symptoms. As long as the majority opinion held that BSE was simply scrapie in cows, fears of a human outbreak were muted. After all, Britons had been eating scrapie-infested meat since the mid-1700s without any apparent ill effects. That clean record suggested that scrapie simply couldn't vault the species barrier between sheep and humans. But even when that was still the majority opinion, years before the BSE Inquiry decided that BSE represented a new prion, some experts were not so sure that humans were safe from infection. In 1988, a young doctor, Tim Holt, and a dietitian, Julie Phillips, published a short article in the *British Medical Journal* in which they categorized the reaction to the announcement of the BSE outbreak as "alarming indifference." And they put together some simple, well-known facts to back up that claim.[1]

They pointed out that BSE was in the same family as kuru and Creutzfeldt-Jakob disease, and that both were transmissible by inoculation, and kuru by eating infected brain. They added that an undetermined number of infected cattle must have gone to the slaughterhouse and been incorporated into the human food supply and, further, that brain could be bought from the butcher raw to put into stew, or already incorporated into meat pies. They then made the prescient comment, "There is no way of telling which cattle are infected until features develop, and if transmission has already occurred to man it might be years before affected individuals succumb."[2]

Years later, Holt denied that they had seen all this in a crystal ball, claiming that "all it needed was a good working knowledge of the literature and common sense. It's extraordinary that more people didn't reach the same conclusion."[3] But perhaps what was even more extraordinary was the reaction to their short paper. One response printed in the same journal, after listing what the writers deemed as weak points in Holt and Phillips's piece, said clearly and bluntly what most medical experts were apparently thinking at the time: "There is, therefore, little reason to believe that bovine spongiform encephalopathy will present any greater threat to humans than scrapie."[4]

Of course, if medical experts were somewhat blasé—or,

more accurately, *unconvinced*—that there was a risk to humans from BSE, you could predict by this point that politicians would be hell-bent on dismissing that risk. Long after the images of cattle carcasses piled up ready for incineration had become common, politicians in England were repeating endlessly that "British beef is safe to eat." In subsequent years, politicians in other countries where BSE surfaced were quick to sing the same tune. They were all wrong.

Mad cow disease was indeed transmissible to humans, and years after BSE was first recognized, the first cases appeared. It was called not mad cow disease in humans but variant Creutzfeldt-Jakob disease, or vCJD, because there were some similarities to the well-known human prion disease CJD but there were also differences. Variant CJD is a grotesque and agonizing condition. It strikes young people, much younger than the typical victims of CJD; it causes dementia, hallucinations, terror, pain . . . the list goes on and on, and the final result, as in all prion diseases, is death. And in a diabolical twist, death comes slower with this new form of CJD.

It was surreal. In England in the mid-1990s, the slowly growing list of young people dying horrible and unexplained (but suspicious) deaths prompted unrelenting reassurances from government scientists and politicians. At the beginning of the decade, the Meat and Livestock Commission had produced a news release titled "British Beef Is Safe." One claim in the release was that "the most eminent and distinguished scientists in Britain and the rest of Europe have concluded there is no evidence of any threat to human health as a result of this animal health problem."[5] That stance didn't waver over the next few years. As late as October 1994, Britain's chief medical officer said there was no link.

Then, a year and a half after that, the health secretary, Stephen Dorrell, was forced to rise in the House of Commons and announce that indeed there *was* a link. At that point, ten cases of this new disease were likely connected to the consumption of meat products from BSE cows.

There is no doubt today. This mysterious new disease of young people is a prion disease, and its prions are virtually indistinguishable from those of BSE. There are now 171 victims in the United Kingdom, though that number peaked in 2000 and has been declining ever since.[6] If you take into account the year the disease peaked, the age of the victims and the date at which the feed bans were put in place, you should be able to estimate how long the mad cow prions have been slowly multiplying inside the victims' bodies. The best estimate is that the incubation period for the disease is about ten years.[7] That means, of course, that many—or most—of these victims were already doomed when the government was claiming that "British beef is safe."

Variant CJD is a tragic story, and although it appears to be coming to an end, with only a handful of cases over the past few years, there could be a whole new surge, a new wave of deaths, and even another after that. And if that weren't bad enough, there could be hundreds of Britons already infected, but unaware, and they could transmit the disease.

In one of the fascinating and totally unanticipated connec-

tions that characterize the science of prions, there are important things about vCJD to be learned from kuru, a disease that
began to go extinct forty years before vCJD first appeared.

As soon as cannibalism stopped among the Fore people
in the late 1950s, the number of cases of kuru began to fall.
Anyone born after 1960 avoided infection altogether (in fact,
there have been zero cases in those born after 1959), so as time
passed, that cohort of kuru-free people grew to be the majority. However, even today, there are Fore individuals who were
alive in the 1950s and did take part in funerary feasts. And the
shocking thing, some of them are *still* succumbing to kuru.

There aren't many left from the 1940s and 1950s, but among
them are individuals who might still die of the disease, apparently having incubated it for forty years or more. John Collinge
of University College London and his colleagues identified
eleven patients with kuru in the years from 1996 to 2004, all of
whom were born before 1950. If each had been infected before
1960, then the minimum incubation time would be forty-four
years, but it is impossible to know exactly when anyone becomes infected; in these cases, it could have been any time in
the 1950s, meaning that the disease could incubate for more
than fifty years.

Can we be sure that these recent kuru deaths are indeed
examples of long incubation periods, triggered by infections
contracted decades ago? Collinge and his team ruled out the
possibility of some sort of later infection caused by either contaminated knives or plates (they were burned immediately
after every feast) or even soil (soil contaminated with scrapie

can be infective for several years, but in New Guinea the feasts did not generate that sort of contamination). They also ruled out the possibility of transmission from mother to child, so there really remains no better explanation than these extraordinary incubation times. This issue of incubation time is crucial, partly for the Fore people who have endured the disease, but also for predicting what might happen in England, where thousands could conceivably be incubating mad cow prions that they ingested sometime in the 1980s.

There is, however, one important difference between the spread of kuru and vCJD. In kuru, human prions were passed to humans; no species barrier had to be breached. In vCJD, on the other hand, cattle prions must cross that barrier to infect humans. There are rules for prion transmission that have been established from both nature and experiments: if a species barrier has to be crossed, the number of animals (or humans) who actually get the disease is reduced, and the incubation times for those individuals are greatly increased.

We won't know what the longest incubation period for kuru is until the last person who could be infected dies. But we know the minimum—that was fixed by the youngest children who got the disease, at four or five years old. From something like four years to forty plus—that is an enormous variation in incubation periods. Now compare that with vCJD: the youngest patients, at the age of twelve, have set the lower limit for the length of the incubation period. Taking just this last comparison, you might be optimistic that there's some sort of proportionality, and if the minimum incubation period is much longer in vCJD than kuru, then the maximum might be as well. On that reasoning, you would hope to see individuals dying of other causes while they're incubating vCJD.

Another way to look at it is to compare average, not minimum or maximum, incubation periods. For kuru, it was about twelve years. No one knows what it is for vCJD because that disease has likely not yet run its course, but it seems that the incubation period for BSE prions in mice is about three to four times that for BSE prions in cattle—again, the difference between a species barrier and no species barrier. Extrapolate those data to kuru and it would suggest that the average incubation period for BSE prions in humans, vCJD, might be thirty to forty years.

There is, however, another key factor: genetics. To understand the role of genes, we have to return to the prion protein molecule. As we've seen, it is, like all proteins, a string of amino acids, each one inserted into its place by a piece of the DNA code, the prion protein gene. Not all prion proteins have exactly the same sequence: different species, and sometimes different individuals within a species, have the occasional amino acid substitution somewhere along the chain. One further complication is that different sequences can be inherited from each parent, so that any individual can have in its body two versions of the prion protein.

It has become clear over the past ten years that amino acid number 129 (out of the 250-plus amino acids in the prion protein) plays a significant role in the chances of getting a prion disease. There are two amino acids that commonly occupy that position, valine and methionine. You can inherit any one of three possible combinations: two methionines, two valines or one of each. There is now solid evidence that having a mix of the two somehow protects against prion disease.[8] Conversely, having identical amino acids in that position predisposes you to prion disease. That applies to sporadic CJD—CJD acquired from human growth hormone, dura mater grafts, you name it.[9] If you're carrying two different amino acids at that position 129—that is, if you are "heterozygous"—you may not be out of the woods, but if you are infected with prions, you'll probably live longer than others, and you might not even get the disease.

Because that protective effect was already known, it came as no surprise to find that the majority of the long-lived kuru victims had this particular heterozygous genetic signature, suggesting that their genes had kept them alive for thirty or forty years. Conversely, their contemporaries who had drawn the genetic short straw had long since died.[10]

Genetics also spawned another truly startling suggestion. One research group pointed out that not only did those who developed kuru late have the right genes but so did an unusually large number of Fore women who had so far evaded kuru outright, even though they had participated in funerary feasts. But that preponderance of heterozygotes among the Fore survivors was not evident in younger Fore people. It seemed as if kuru was acting as an evolutionary selective force, allowing only those with the right genes to survive. That makes sense. But the remarkable thing was that when the scientists compared these genetic results with the genes of other groups of people around the world, they found evidence of the same thing: a preponderance of that mix of two different amino acids at site 129 in the gene for the prion protein. If, as they suggested, there were more of these than you'd expect, you'd have to conclude that there was some sort of selective pressure acting not just on the Fore but on all these populations in different parts of the world. What could that pressure be?

As far as the researchers were concerned, it could only have been cannibalism: operating worldwide, over long periods, millennia ago, acting to circulate some unknown prehistoric prion. Given prevalent attitudes toward cannibalism, discussed in Chapter 3, it's no surprise that many found this idea, well, distasteful. That also explains why it made headlines everywhere. Far fewer headlines were devoted to a study a couple of years later that argued that the genetic analysis was wrong and that there was no hard evidence of any selection, no unusual distribution of prion protein genetic types, and therefore no conclusion about worldwide cannibalism to be drawn.[11]

Let's face it: while speculation that cannibalism was rampant in the world long ago is intriguing, it wouldn't change life today. But there remained another side to the discovery that kuru was still killing people. And it has important—and chilling—implications for everyone.

The evidence that heterozygous Fore women were protected against kuru (either completely or at least over the short term) is inarguable. It is also indisputable that heterozygosity—both methionine and valine at position 129 in the prion protein gene—is apparently protective against all prion diseases, or at least those that strike humans. With that in mind, here are the facts about vCJD. There have been 171 cases in England and a couple of dozen in other countries. All English cases were homozygous; that is, these individuals had inherited the gene for the amino acid methionine at position 129 from both parents. Their genetic makeup was m/m. Surveys had already shown that m/m individuals make up about 37 percent of the British population.

Consider this: 171 cases so far, all drawn from about a third of the population, a third that can be considered to be the most susceptible to prion disease. What about the rest, who have different genes? Another 13 percent are thought to be valine/valine homozygous, and the remainder, about 50 percent, are heterozygous. The implication is pretty straightforward. Yes, people have contracted the human version of BSE from eating meat products, but the ones who have are—sadly—those you'd expect to be first in line because of their genetics. The decreasing incidence of vCJD may simply represent the fact that this first, most vulnerable, cohort has been exhausted. They might have been the most vulnerable, or they might be those with the shortest incubation period—both are true of that homozygous condition.

That still leaves two-thirds of the British population. How come none of them, the other two genetic types, have died of vCJD? Maybe they escaped infection. That's possible, but it's also possible that they *have been* infected and are living through a longer incubation period, one that will nonetheless end with disease and death.

It is impossible to know absolutely, but researchers in England have done an incredible job of working with scant data to try to predict just how many people in the United Kingdom will die of vCJD. The changes in thinking about a

future epidemic are described best by the estimates, and how,
as time passed, they have changed.

Changeable, and none too precise either. For instance, in
1997, working with the scant data available then (the existence
of vCJD had only been *acknowledged* the year before), one
team came up with estimates of final casualties ranging from
a few hundred to tens of thousands. How could they do any
better? They didn't know the most fundamental things about
the disease, such as who had it, how long had they had it and
whether they could spread it.

So here are some other estimates of the final number of
cases of vCJD and when they were made:

1997: a few hundred to a few thousand

1999: a hundred or less

2001: anywhere from a few hundred to several million
(although most of those millions would die of something
else before they died of vCJD, meaning that actual epidemics would "only" be several thousand cases)

2002: 7,200 infected

2003: anywhere from 10 to 7,000.

It's tempting to laugh—or sigh—at what looks like statistical
floundering, but these are the very best analysts trying, under
incredible pressure, to come up with an answer when an answer isn't available.

It is agonizing, really. It could be that there are hundreds, if not thousands, of people in England who seem perfectly healthy but arc incubating prions that might one day kill them. It's also possible that, because of their genes, their personal incubation periods will turn out to be extraordinarily long, long enough that they will die of something else before they die of human mad cow disease. It's even possible that their genes have protected them from infection altogether. Obviously, it is important to be able to tell which of these options is likely to be true, crucially important for the society in which they live.

Take the first possibility, that there are people in the incubation phase of vCJD and they don't know it. If that's true, then the possibility looms of three waves of vCJD. The first, which struck people of m/m genetic type, is just about over and claimed (at last count) 171 victims. The second, which apparently hasn't started yet, if indeed it will at all, would be expected to be the other homozygous condition, valine/valine, and would be smaller in scale (because there are fewer of them).[12] It wouldn't surprise anybody if a second wave was starting. Of course, even if it were, that wouldn't mean that the first wave was over. They might overlap for years.

But here's why these thoughts are so vitally important: no matter what their genetics, people walking around incubating vCJD are potential transmitters of the disease. It's been said that it's not the cows we should fear, it's the people. Human-to-human transmission eerily reminiscent of kuru.

And this isn't speculation. There is already hard evidence it's happened. At least three people in England have contracted vCJD from blood transfusions, infected by donors who, unknowingly, were incubating vCJD. When these cases were investigated, it was discovered that of sixty-six people who had received blood from donors who then went on to develop vCJD, thirty-four died too early to have developed the disease. Of the thirty-two who were left, twenty-four were still alive. There had been eight deaths: three of those had had prion infections, and two had died of prion disease.[13] That is scary enough, but in this case, one of the donors didn't develop the disease for a further twenty months; another infected donor for a further forty months. They could theoretically have given blood several times over that period (these two didn't, but you'd have to think that others have).

Even these small numbers have their revelations: the minimum incubation by this route turned out to be about six years. That's much more rapid than getting vCJD from eating meat contaminated with BSE. The transmission is more efficient because it's intravenous and also because it's species to species, with no barrier to cross.

A recent study came to the conclusion that while infection through blood transfusion has happened, it will never create a lasting epidemic.[14] That is partly because now all white blood cells are removed from donated blood, a process called leukodepletion, which reduces infectivity by something like 40 percent. Also because anyone in the United Kingdom who received a blood transfusion later than 1980 isn't allowed to donate. This study also pointed out that prions transmitting this way are on a one-way street: most donors are young; most recipients are old, so prions move from the young to the old, but they can't backtrack. Most young people aren't blood recipients; most older people aren't donors.

But overall, more than a decade after the first cases of vCJD were announced, there is still uncertainty. How many people are carrying vCJD prions and don't know it? What is the risk that those people, however many there are, might infect others? And are we looking at the possibility of future epidemics based on human-to-human transmission? There are some answers to these questions, all of them tentative.

First, how many are infected? There is not yet a rapid blood test for the presence of prions, so one of the only ways of estimating these numbers is to look for prions in tissues that might harbor them. Fortunately (a word that's rare in this business), vCJD is quite different from sporadic CJD in terms of the tissues it infects. Variant CJD prions are found in tissues like the tonsils, appendix and spleen long before they enter the nervous system and cause detectable disease. (In fact, they might remain there for the lifetime of the individual, never causing death.) This appearance of the disease prions in these tissues is not seen in sporadic CJD.

Appendixes and tonsils are often removed surgically, which opens up the possibility of examining those discarded tissues for the presence of prions. This has already been done once in the United Kingdom on a grand scale.[15] An analysis of 11,109 appendixes and 1,565 tonsils turned up three samples with vCJD prions in them.[16] Two of those were tested genetically and shown to be homozygous for valine, that genotype of which only one case of disease has just been found. What conclusions can be drawn from these numbers? In one sense, they're reassuring: a 0.02 percent rate of discovered infection seems pretty small, although if you scale that up to the current population of the United Kingdom, about sixty-two million, it would suggest there might be more than twelve thousand people infected but not yet showing symptoms of disease. Add to that the possibility that, early in infection, there are not enough prions to be detected by lab tests—which would imply that the three infected cases is an underestimation—and this study is justifiably worrying.

It's not just the possibility that people incubating prions might give blood that's frightening; it's also true (as has been known since the tragic accidental infections of the 1970s) that surgical instruments can be conveyances of infection. Prions have a powerful affinity for stainless steel, and the standard methods for sterilizing surgical tools are not equal to the job of removing them, especially if they've gone through one inadequate cycle of heat sterilization and are baked onto the surface of the instrument. The risks depend on two key factors: one, the ease with which surgical instruments pick up prions from an infected brain and then transmit them to a second patient, and two, the number of times a surgical instrument is reused.

It's hard to avoid contamination of an instrument from an infected brain and just as hard to estimate how easy subsequent transmission is. But you might think that reuse would be dead easy to control: just don't use any instrument more than once. Yet that's easier said than done. Single-use tonsillectomy tools, for example, were introduced in the United Kingdom, but it was soon discovered that postsurgical complications, including hemorrhage, were significantly higher and their use was discontinued. These risks were deemed to be higher than the risk of getting vCJD from the tools.

In the United Kingdom, the current strategy to prevent further transmission of vCJD through blood transfusions is to prevent anyone who has received a transfusion since 1980 from giving blood themselves. In Canada, you can't donate if you've received blood in the United Kingdom, France or western Europe since 1980, or even if you've spent a total of three months or more in the United Kingdom or France (which has its own vCJD problem) between 1980 and 1996. The United States has similar restrictions.

And that is where we stand today. If you spent significant time in the United Kingdom (or France) in the 1980s and ate meat products, it's conceivable that you are incubating prions. It is probably not very likely, but no one can give you complete reassurance that you are not. Will dozens more come down with vCJD? Hundreds? Thousands? Who knows? This disease and these prions are just too new, too poorly understood. At least for now.

16

The Americas
Mad Mink, Then Cows

For obvious and very good reasons, the world has focused on the prion diseases of Great Britain, but North America is home to a couple of its own very strange varieties. One is important by virtue of its past; the other by its future.

The first is something called transmissible mink encephalopathy (TME), a disease of ranched mink. It is not at all common, but it might be significant. The earliest recorded case was in 1947. Mink living on a ranch in Wisconsin came down with a weird neurological disease: they became even more aggressive than usual (mink are rarely friendly). They arched their tails over their backs like angry squirrels, attacked the wire of their cages and inevitably died, sometimes with their teeth still clamped around the wire. This was followed by other, isolated incidents of the same kind at different mink ranches. The first good guesses, years later, suggested that it was a scrapie-type disease (in the mid-1960s, scrapie was really the only animal disease that could serve as a model). But the scrapie theory ran into problems: if you inoculated scrapie directly into the minks' brains, the animals did indeed become diseased; but nothing happened if the mink *ate* scrapie-infested material, which had been the suspected route of infection: mink getting the disease by eating processed food laden with bits of scrapie-ridden sheep. This was a puzzler, because other than scrapie there really weren't any good candidates for an infectious agent.

Then there was another outbreak, this time in Stetsonville, Wisconsin, in 1985. This incident was the most important yet because the owner kept meticulous records and swore that he had never fed sheep to his mink—only "downer" cows. As the name implies, these are cows that simply can't get up or stay on their feet, and they're culled immediately. In such cases, the disease responsible for the cow's immobility may never be determined, simply because there are several conditions that present this way. Estimates of the number of downer cattle in the United States every year range wildly, from less than one hundred thousand to as many as seven hundred thousand. With today's wisdom you could certainly suspect that some downer cows actually had mad cow disease and could serve as a source of infection. The thing is, if you weren't looking for mad cow disease—and you wouldn't have been if you were in Wisconsin in the mid-1980s—you wouldn't see it.

If the Stetsonville mink really hadn't consumed sheep remains, then it couldn't be a mink version of scrapie that was killing them. But they had had their fill of processed cattle. That wasn't much to go on, especially considering that at the time there was no evidence anywhere that cattle could be infected by a scrapie-like disease (the first cases of what would become mad cow disease were just surfacing in the United Kingdom). Richard Marsh, the undisputed expert in TME, wondered whether whatever was causing this disease—and it looked like one of those newly named prions—could move between cattle and mink. If it could, and relatively easily, that might mean it was acclimated to both—maybe had even spent some time *in* both. So Marsh inoculated two steers with material from the mink. They became ill within two years. When samples of their brains were inoculated back to mink, the mink too became ill. It didn't prove that mink were dying because of a cow prion, but it certainly lent support to that idea.[1]

There had also been a much earlier case—in Ontario, Canada, in 1963 (a year when there were also two outbreaks in Wisconsin and one in Idaho)—where the owner was positive he hadn't fed sheep to his mink. Bill Hadlow, the man who made the brilliant connection between kuru and scrapie, wrote an article about the Ontario case in which he described how there was no evidence of any consumption of sheep (at the same time acknowledging the possibility that sometimes, unbeknown to the rancher, there were sheep remnants thrown in with the beef) and pointed out that there was evidence already that scrapie wasn't the responsible agent for these mink outbreaks.[2] An example: you could inoculate goats with ground-up mink brains and they would get a disease all right, but it was totally unlike scrapie in goats. Two completely different disease profiles meant two different diseases.

Marsh insisted that there must be a cow disease in North America that had infected the Stetsonville mink. If there were, it would be North America's own version of mad cow disease. For obvious reasons, this didn't go over very well in his native United States. Marsh had evidence that his critics would have a lot of trouble explaining away, but criticize they did anyway,

even when compelled to be somewhat civil, as in this note in a
publication of the Animal and Plant Health Inspection Service
(APHIS) of the U.S. Department of Agriculture:

> Although Marsh's hypothesis is based on speculation and
> anecdotal evidence, in 1993 APHIS adjusted its national
> BSE surveillance program to include testing downer cat-
> tle for evidence of a TSE [transmissible spongiform en-
> cephalopathy]. The brains of more than 20,141 cattle have
> been examined at APHIS' National Veterinary Services
> Laboratories and other State diagnostic laboratories. Not
> a single tissue sample has revealed evidence of BSE or
> another TSE in cattle.[3]

Not, "We've not found any evidence yet," or "To date there's
no evidence of a TSE," but that dismissive "not a single tissue
sample."

Marsh was claiming the existence of a prion disease of cows
in North America at a time when there hadn't been a single
case of BSE discovered there. (That has changed since.) But
BSE was now becoming explosive in the United Kingdom, and
Marsh's critics were undoubtedly determined to prove him
wrong. But he hasn't been proven wrong, and until he is, the
possibility exists that there is some sort of sporadic BSE-like
infection of cows out there, a prion that is infectious when
it's included in meat and bonemeal. Funny that it didn't catch
fire in cattle, though.

And what about cattle? When the first case of mad cow disease was found in Canada, it ruined the Canadian beef indus-

try, as all world markets abruptly stopped importing Canadian beef (although many of those markets have since reopened, believing the human risk to be negligible). Since that first case in 1993, there have now been seventeen more in Canada. The United States has reported two of its own, and an additional one that had been imported from Canada.

The eighteen Canadian cases have all been attributed to contaminated feed. Case number 7—a dairy cow on a farm in northern Alberta in the summer of 2006—is a good example of how you can arrive at that conclusion. Case number 7 had apparently died of complications from an infection, but because she was a downer cow, she automatically qualified for a BSE test. The test was positive. Besides identifying all the other cows that had had access to the same feed, *and* all the cows born into that herd, the feed itself had to be accounted for, this despite that in 1997 Canada established a feed ban prohibiting the inclusion of any ruminant waste into feed that would be given back to ruminants.

Why worry about the feed as a possible source of infection if there was a ban in place? Because, as had been discovered in the United Kingdom, enacting a ban is one thing; making it work another. Thousands of cattle with BSE had been born there after that country's ban, either because of noncompliance or cross-contamination of cattle feed by feed designated for pigs or chickens. Most of the born-after-the-ban cows in the United Kingdom, at least initially, tended to reside on farms where there were lots of chickens and pigs being fed meat and bonemeal.

Case number 7 was born in 2002, several years *after* the Canadian ban on meat and bonemeal. Contamination immediately leapt to mind, but the Canadian Food Inspection Agency's analysis showed that the contamination of cattle feed by food products for other animals on the farm seemed not to be an issue. Feed for the hens and rabbits on the farm did not contain prohibited material; horses and goats were not fed commercial products at all. Prohibited material was fed to the cats and dogs on the farm but not to the cows.

For the first two months of its life, case number 7 ate nothing but feed from facilities that contained no prohibited material. As it matured, it was fed a mix of prepared feed from three different suppliers. One had been free of prohibited material for more than ten months; the other two were set up to comply with the 1997 ban, even though they were receiving material from a renderer who had been implicated in previous cases of BSE.

The only slender hint that something might have gone wrong was that one of these companies did not document that the equipment used to produce cattle feed had been flushed after producing feed for other animals that weren't subject to the ban. Did it flush the equipment? We'll never know, but given the incredible hardiness of prions, failing to flush machinery could easily allow remnant prions to survive and contaminate the next load of feed.

The question is, when will it seem unreasonable to continue to point to possible contamination? Even the latest case (as of writing), a dairy cow in Alberta discovered to have BSE in February 2011, was six and a half years old and so could conceivably have been fed from bins contaminated with meat and bonemeal. In mid-2007, the ban was extended, so that something called SRM, specified risk material, could not be included in any animal feed, pet food or fertilizer. This includes part of the cow's intestine, the brain, eyes, tonsils, spinal cord and other tissues thought to harbor prions. Calves born from now on, provided there is compliance in Canada, should be BSE-free. If they aren't, maybe the scientists will have to admit that there are other sources for BSE infections. And why shouldn't there be? There are sporadic cases of other prion disease, like the human CJD—cases where there appears to be no known cause. If that happens, people will remember Richard Marsh and his claims for the Stetsonville case of TME.[4]

It is somewhat reassuring that the worldwide incidence of BSE has totally run out of steam. Feed is the key: fix that and the epidemic starts to die. It's not dead yet—and might never be completely—but it's not as scary as it was in the mid-1980s.

17

Into the Wild
Deer, Elk, Moose and Caribou

At the moment, there is only one prion disease in the world that is spreading. It now represents the largest biomass of prions in the world. It is also the only one that has infected both captive and wild populations of animals. It is unclear how it spreads. And it is uncertain how *far* it will spread. One thing is sure: prion scientists are very worried about it.

It started in 1967 at a mule deer research facility near Fort Collins, Colorado. The deer were losing weight, salivating heavily, drinking a lot, alternating between listlessness and hyperexcitability, pacing in that familiar pattern seen in zoo animals and eventually wasting away or dying suddenly after being handled. Usually death came within a year. At first it was dismissed as some sort of negative reaction to confinement—"ain't doing right" was the way it was described. But it soon became clear that this was a unique disease syndrome. Researchers named it chronic wasting disease, or CWD. Then it began to spread, first to captive deer in Wyoming, then to captive elk. In the 1980s, the first *wild* animals carrying the infection were found: elk in Wyoming, more elk in Colorado, then mule and white-tailed deer in both states. Most scientists agree that CWD, although only noticed in the 1980s, was probably around in these wild populations—unremarked—at least twenty years before that. Then, the more researchers looked, the more they found. Likely as a combination of both greater attention being paid and actual spread, CWD can now be found in eighteen states and two Canadian provinces, in both captive and wild animals.

The spread is steady if not spectacular, but there is still much that isn't known about CWD. For instance, most of the wild animals found by surveillance to be incubating the disease are not yet even showing symptoms; probably by the time they do, they easily fall prey to predators, then scavengers, and are never seen by a human. So the numbers reported are only a rough estimate of the actual incidence of the disease.

There is a worrying precedent in prion science for this. During the BSE epidemic in Britain, even though the number of sick animals was tracked closely from the very beginning, nothing was done as long as the disease seemed intermittent and sporadic. By the time it was clear that it was indeed an epidemic and action was taken, much damage had already been incurred. It's the same thing in North America: nobody really knows how many deer or elk have died, are sick or are incubating CWD.

How is it spreading? This is not a replica of BSE: those wild animals with CWD are not being fed anything that could perpetuate a prion infection. But infected animals do pass the disease to healthy ones, although exactly how this happens is still a mystery. It could be that an infected animal nuzzles others—there are plenty of prions in the saliva, and these are passed on by animals that are not yet exhibiting any symptoms of the disease. But it's also probable that urine, feces and placentas of infected animals are contaminating the landscape. And that could be a huge problem. These prions, like those of scrapie, remain infectious for incredibly long periods in the natural environment. Just as farmers found out hundreds of years ago that introducing sheep to fields where scrapie had existed years before risked causing new infections, researchers have noted that mule deer are infected by grazing in pens where CWD deer had lived two years earlier. Also, mule deer that came into contact with either the bones of CWD-positive dead deer or the surrounding vegetation were also at risk for CWD. Given the ease of infection, it's no wonder that rates of disease are extremely high among captive populations. They are, of course, lower in wild animals, which are constantly on the move.

"On the move" might be understating it, though, and this too is a complicating factor. While it is true that most adult deer don't stray more than a few miles from season to season, young deer have been shown to wander much further. One female mule deer, tracked by radio collar, traveled more than 300 miles in a year and a half, and at one point was nearly 60 miles from where she was born.[1] Had she been CWD-positive and infectious through her saliva, she could have brought the disease to entirely new areas.

Even with this apparent potential for spread, it might be tempting to dismiss the threat of CWD, given that it exists in nothing like the density of mad cow disease in England; yes, many animals are infected, but they are spread over vast areas of North America. But the beginnings of significant events often seem inconsequential, as did the mad cow epidemic in the early stages.

Consider this study of the mule deer population in the Table Mesa area southwest of Boulder, Colorado.[2] Their home is not the wilds of the deep forest: these deer move in and around a combination of woods and suburban housing. Mike Miller, a veterinarian with the state of Colorado, and his colleagues tagged 131 of the deer and followed them over time. They found that about 25 percent of them were infected with CWD, and an additional 25 percent became infected each year of the study. That is an unusually high rate of infection compared with other areas; CWD has been there since 1985, and the mule deer population is now about half the size it was. The life expectancy of infected deer was reduced from more than five years to less than two, and predation by mountain lions on those deer increased fourfold, even though few of the deer killed by mountain lions appeared ill prior to death.[3] In fact, the high rate of mountain lion attacks might be exacerbating the situation because deer tend to cluster together more intensely in the face of heightened predation and enhance transmission of the disease by doing so.

But reaction to predator pressure is only one aspect of the natural behavior of deer that likely plays an important role in the dynamics of chronic wasting disease. Deer, elk and moose move to higher ground in the summer and return to the valleys in winter; the young males disperse from their original group; and, of course, there's that odd case of the nomadic female—all of which conspire to make the spread of CWD complex and mysterious. But the Table Mesa data alone should dissuade anyone from thinking this is a trivial issue.

Anyway, the spread of CWD need not depend on the natural movements of wild animals. Shipping captive deer and elk across state lines, and even from country to country, has helped spread the disease. New York State's CWD cases first appeared in captive animals imported from other states, then spread to wild deer. Even South Korea has had cases of CWD. The animals were imported because of the Asian fascination with antler velvet, yet another substance thought to enhance male fertility—*human* male fertility, that is.

These two features—the fact that CWD is the only prion disease infecting both domestic and wild animals *and* that it is spreading in mysterious ways—have put the spotlight on CWD. The problem is not simply that the disease is a threat to deer, elk and even moose (although that should be enough) but that it might infect other wild animals, and that it might even be transmissible to us. Sounds like fear-mongering? Well, there are thousands of people who eat venison and elk in the United States and Canada. Eating cows with BSE has proven fatal for humans. And, unlike BSE—and more unsettling—the CWD prions have been found in the muscle tissue of mule deer. And when material from muscles was inoculated into mice, the mice became ill, although it took much longer than when they were inoculated with material from infected brains. A longer incubation period like that suggests that levels are lower in muscle than in brain, but nonetheless, the authors of that study claimed that deer muscle is "a major source of prion infectivity. Humans consuming or handling meat from CWD-infected deer are therefore at risk to prion exposure."[4] Note that "exposure" doesn't necessarily mean infection.

How would we know if people were dying from consuming

prions from deer or elk infected with CWD? First, we'd have to assume that the symptoms would be something like the other human prion diseases, especially sporadic Creutzfeldt-Jakob disease, or its diabolical mad cow–related human version, vCJD. With that in mind, it seemed that in the late 1990s that an investigation was launched any time someone who had eaten venison died of a neurological disease. And there's no doubt that some of these deaths looked suspicious. For instance, one study focused on three individuals who died of Creutzfeldt-Jakob disease at a young age: two had been hunters and the third was the daughter of a hunter. Two of the patients were twenty-eight when they died; the other was thirty—all very young for CJD (which mostly strikes people over fifty-five), but, of course, reminiscent of vCJD. All had eaten venison regularly.

Suspicious, yes, but there were serious inconsistencies. One patient, the hunter's daughter, had eaten deer meat. all right, but it was from Maine, where no CWD has ever been found. She had had elk meat from Wyoming too, but only twice, when she was six years old.

The second patient had eaten deer and elk usually once a month, and those animals had mostly been hunted in Utah, a state with very little CWD. The third patient had eaten ground venison, but as far as his wife could remember, none of it had originated in either Colorado or Wyoming, the two states with the most infected animals.

These individuals represented three of only four in all of the United States who had died of CJD at or under thirty years of age from 1996 to 2000. The fourth had had no contact at all with venison or elk meat. Not only was there a disconnect between the source of the meat and areas where CWD is common, but the brain pathology of the victims made the link to CWD doubtful as well. It was typical of sporadic CJD, and totally unlike that of variant CJD, which is, of course, the only example we have, besides kuru, of transmission of a prion disease by eating contaminated tissue. At the same time, we don't really know what a case of CWD in humans would look like—the best that can be done at this point is to assume there's some resemblance to other human prion diseases.

Could these cases simply have been coincidental? Plenty of people in the United States eat deer and elk, and even given the rarity of CJD—one in a million—three cases wouldn't be impossible. Other studies have shown that the overall rate of CJD in counties in Colorado where CWD is endemic was no higher than in counties where the disease is absent, nor was the rate of human disease on the rise over the last two decades of the twentieth century.

Yet, it's impossible to dismiss the possibility of a human case of CWD. What about that elk meat the woman ate when she was six? Although it was supposed to have come from Wyoming, the young woman's mother had only a vague recollection of it, and it proved impossible to trace. One of the other patients hunted an elk in southwestern Wyoming in 1995, but had it been diseased, the incubation period—the time from eating the meat to coming down with the disease—would have been only a couple of years, and that seems too short to attribute the disease to the consumption of infected meat.

But these four cases aren't the whole story: there is, at the moment, a large-scale unintended experiment under way con-

cerning the risk of disease from eating infected venison. In
March 2005, the Verona Fire Department in Oneida County,
New York, hosted its annual Sportsmen's Feast, at which was
served, among other things, venison in the form of steaks,
chili, stew, sausage and meat patties. After the banquet, tests
revealed that one of the deer that had been part of the mix was
positive for CWD (making it one of the very few deer with
CWD in the entire state). Although no organs from the deer
had been served, participants had presumably been exposed
to the CWD prions, not just by consuming the meat but also
by butchering and cooking it and coming into contact with
contaminated surfaces.

Eighty-one of the attendees at the feast thus became the
unintended participants in a follow-up study of the risks of
consuming CWD-contaminated meat. The group was not all
equally exposed: fifty-six had only eaten the venison, whereas
seventeen had eaten the meat and also participated in its preparation. Of course, those were risks associated only with the
single infected deer that had been the uninvited guest at the
banquet. Given that it was a Sportsmen's Feast, many of the attendees were hunters and had killed, butchered, dressed and
eaten other deer in the months and years before this incident.
Those risks too have been tabulated, and now the waiting has
begun. The study, which will monitor the health of those involved with careful cross-referencing to cases of what appear
to be CJD in that population, is scheduled to continue until at
least 2014.[5]

Researchers have tried to determine in the lab if humans
could be susceptible to CWD prions, and they have come up

with mixed results. One study in 2010 showed that CWD prions injected into the brains of mice engineered to have human-type prion proteins had no effect. But the researchers were cautious, allowing that they might not have been using the most appropriate strain of CWD.

Maybe they were right. At the beginning of 2011, a startling paper was published that argued if the correct lab techniques were employed, there was a way of infecting humans, or at least of causing human proteins in cell culture to misfold.[6] In this case, the researchers used a technology that accelerates prion replication dramatically. When CWD prions were put through several cycles of reproduction with deer prion proteins and the resultant mix was exposed to human prion proteins, there was some misfolding. Take this at face value and you'd have to conclude that some variants of CWD, under the right conditions, might cause disease in humans. But the experiment is highly technical and therefore at some remove from what happens in nature, so any interpretation at all has to be tentative.

But you don't have to be killed by a prion disease to be hurt by one. The impact of finding BSE in Canada has had a devastating impact on the beef industry and especially on farm families. Mad cow is a very different disease from chronic wasting, but there are worrying scenarios. For one thing, it seems inevitable that deer with the disease will cross paths with cattle. Could CWD prions be lethal to cattle too? That question is a difficult one: it is true that if cattle have CWD prions inoculated directly into their brains, they come down with disease. But oddly enough, this transmission is much more efficient using brain tissue from white-tailed deer rather than mule deer—there is something going on at the molecular level that differentiates the two. And anyway, that doesn't prove that infection could occur in a natural setting, where cattle would be somehow picking up prions that deer left behind in the environment.

The risk to cattle is one issue, but there is another more ominous—but at the same time extremely uncertain—risk and that is to another species altogether: caribou. Chronic wasting disease has spread substantially north and east from where it began in the American west in the 1960s. It is now well established in Saskatchewan and Alberta and is gradually working its way north. And don't forget: the number of deer that actually die of CWD may not give you an accurate picture of how far the disease has spread because there are many incubating the disease, and we don't really know where they all are.

Deer are randomly killed and analyzed, so there *is* surveillance, and it has uncovered some curious facts. For instance, it appears, at least in Alberta and Saskatchewan, that mule deer are much more vulnerable than white-tailed deer—by a ratio of about ten to one. Infected deer move along river valleys and in rough terrain, but beyond this, there is very little certainty. Regardless of the actual pace, CWD is migrating north, and there is particular concern about Saskatchewan. The northern part of that province is home, at least part-time, to herds of both woodland and barren-ground caribou, tens or even hundreds of thousands of animals. These caribou undertake one of the largest annual mass migrations anywhere in the world. In turn, Aboriginal groups in northern Saskatchewan depend on the annual arrival of the herds. It's been estimated that in meat value alone, the northern caribou herds are worth $100 million, but their cultural value to the people who have lived with them for centuries is beyond any dollar value.

So the shadow of a threat looms: if infected deer bring CWD far enough north to intersect with the caribou migration routes, and if caribou are susceptible to CWD, then we have the makings of a disaster. As far as part one goes, infected deer are being found farther and farther north. Part two is the key: Are caribou susceptible? As I was finishing this chapter, news came of the completion of a study showing that, yes indeed, CWD prions from white-tailed deer could infect caribou—but maybe only some. Of six animals, half fed deer prions, half fed prions from elk, two of the three caribou infected from deer caught CWD. The third animal, and all three infected with prions from elk, were unaffected.[7] The results are not completely surprising because there is evidence that the prion proteins in deer are more similar than their counterparts in elk to those in caribou. That might make the caribou prion proteins more likely to misfold in their presence.[8]

But uncertainty about CWD will continue. In this latest experiment, the one caribou which seems to have been resistant to infection from deer was subtly different genetically. If that genetic difference is indeed protective, then the question is, how widespread is that "safe" genotype in the vast herds of caribou? What can be said without doubt is that prion scientists are worried about its potential scope; some even think it could in the long run threaten the entire North America population of the cervids: deer, elk, moose and, perhaps most dramatically, caribou. For now, it seems like a remote possibility, but then, so did BSE.

18

Origins
Attempting to Find Where Prions Come From

I've been selective in this book: the prion diseases I've described in detail are those that have had the greatest impact from a scientific, public health or natural history point of view. There are others: fatal familial insomnia and Gerstmann-Sträussler-Scheinker in humans; atypical versions of BSE and scrapie in animals. There's little doubt that in the future new prion diseases will surface or that diseases that until now have been difficult to categorize will turn out to be prion diseases. But even with the enormous amount of data that have been gathered about these weird infectious agents and the invariably fatal diseases they cause, one question can still be asked of each one: Where did they *come* from? It's an important question, not just to satisfy scientific curiosity but because in some cases the answer could shed light on the sorts of risks we might see in the future. And so, in no particular order, some thoughts on the origins of prion diseases.

Kuru

Kuru is a perfect example. The Fore people testified that kuru didn't appear until the early part of the twentieth century.

Then, of course, over no more than three or four decades, it spread dramatically, killing two hundred people annually at its peak. In this case, it's known how the disease was spread but not how the index case arose. It looks like it was a piece of incredibly bad luck: that right around the time the Fore began to have funerary feasts, one of them happened to die of CJD, the sporadic kind that affects one out of every million people. They might never have seen CJD before but, regardless, that individual was consumed and the disease began. This would make kuru a kind of CJD transmitted by consumption. Kuru is the only example of this in humans.

Yes, it could have happened this way, and yes, the prions from the two diseases are pretty much alike, but they cause different symptoms and target different areas of the brain. Victims of kuru were most often still mentally intact until just before death but were incapable of movement: the prions had virtually destroyed their cerebellum, the lump of brain tissue at the back of the skull that controls and coordinates movement. But CJD sufferers become demented early on in the disease, a result of more widely distributed brain lesions. Some of these differences could be attributed to the route of infection, kuru being oral, sporadic CJD being, well, who knows? It is not supposed to be infectious at all. Where *does* it come from?

Creutzfeldt-Jakob Disease (CJD)

One person in every million, worldwide, comes down with what is known as sporadic CJD every year. But is it truly sporadic, arising suddenly and unpredictably in an otherwise smoothly functioning brain? And what exactly would *sporadic* mean in molecular terms? For instance, there are prion diseases in which victims are born with a mutation in their prion gene, a change that presumably makes their prion proteins more susceptible to misfolding. This predisposition suddenly makes its presence felt, usually in middle age, crossing some sort of critical threshold to create full-blown disease. These diseases include fatal familial insomnia and Gerstmann-Sträussler-Scheinker disease. In both cases, a single mutation, a single amino acid substitution in the prion gene, invariably leads to the disease. There are also familial versions of CJD, distinguished from the sporadic form of the disease by the presence of a genetic trigger. Some experts believe that drawing hard and fast lines between these different prion diseases is a mistake, that they are, in the end, an entangled mess of the same disease with different faces.

But the familial versions exclude sporadic CJD, that version with the well-attested one-in-a-million incidence.[1] Are all these cases truly sporadic, appearing out of the biological nowhere? It would make more sense if they were in fact the result of infection by some unseen and unknown prion, but there's almost no evidence for infection, or for the transmission of the disease from one individual to another. I say "almost" because of one remarkable experiment. Mice were genetically engineered to have human prion genes, making them, in a sense, like us: the prion proteins on their nerve cells were the human, not the mouse, version. These humanized mice were then infected with BSE. The vast majority of them came down with a disease similar to variant CJD, which makes sense and was expected because it echoes the real-world experience of people in the United Kingdom who ate infected beef products. But that wasn't the whole story. The prions from a few of these mice seemed much more similar to those from sporadic, not variant, CJD, and the brain of one of the animals had sustained damage much more like sporadic CJD than variant CJD.

But that's not supposed to happen! Sporadic CJD is not believed to be the result of any kind of infection. This experiment suggested that perhaps some of those "sporadic" cases—at least recently—might actually have resulted from infection by BSE. It was interesting in this regard that Switzerland, which had the highest rate of BSE outside England, experienced a doubling of sporadic CJD in 2001. But no one was able to turn up any convincing evidence that the two diseases were related. And at the same time, there are solid arguments why most cases of CJD have nothing to do with BSE or any other prion disease. CJD existed long before BSE did; most of its victims live far away from the United Kingdom and have never visited the United Kingdom. Could scrapie be a candidate for a secret prion causing CJD in humans? Unlikely, given that Australia and New Zealand are free of scrapie and yet both have sporadic CJD. So unless there's some other prion circulating among us that occasionally triggers disease (and that would be *extremely* hard to find), sporadic CJD is just that—random and unpredictable.

Before leaving sporadic CJD, here's another thought. A few years ago, Italian scientists discovered that there's a second kind of BSE, a sort of mad cow II. This alternative prion disease in cattle is called BASE, the "A" standing for "amyloidotic," referring to the presence of deposits of misfolded proteins called amyloid. These were elderly cows, and one of the unsettling things about them is that even though their brains bore signs of this new prion disease, they hadn't shown any *symptoms* of mad cow, meaning, of course, that cows with this version of the disease would go to the slaughterhouse with no one knowing they were loaded with prions. But there was another, even more curious, thing about these cows: the molecular characteristics of this new prion were more like the agents sometimes found in sporadic CJD than were the prions of mad cow. Both also created brown aggregated deposits in the brain. If you lean toward the idea that at least a few occurrences of CJD might have some sort of infectious origin, this is interesting stuff, but it's very, very speculative, there aren't nearly enough data yet, and there are still plenty of questions to answer before we start worrying about this. How common is this atypical BSE (it seems pretty rare)? Is there any geographic/temporal overlap between cases like this, especially in older cows, and the occasional clusters of sporadic CJD? These are good questions, good enough to warrant trying to find the answers.[2]

And those clusters: even though sporadic CJD runs at about one case per million people every year, it isn't distributed evenly. Just as you get runs of heads or tails when flipping a coin, there are clusters of CJD cases. The thing is, in the case of the coins, it's just randomness; in the case of CJD, we just don't know.

For instance, in 2004, the U.S. Centers for Disease Control and Prevention (CDC) investigated the case of seventeen people who had gone to the Garden State Racetrack in Cherry Hill, New Jersey, sometime between 1988 and 1992, and who

had then died of CJD over the next several years. All were over
fifty years old. There were fears that somehow all these people had eaten contaminated meat at the racetrack restaurant
in the late 1980s or early 1990s. But a closer look changed the
picture: first, three of the deaths were diagnosed as conditions
other than CJD, reducing the list to fourteen. Then the CDC
did the math: more than four million people had been to the
track in those years, of whom approximately three hundred
thousand would have been over fifty; while the CJD death rate
for the entire population is one in a million, for people over
fifty it's between three and four per million per year, and so
fourteen deaths is within the limits of probability. The issue
was complicated by the fact that there were multiple food suppliers servicing the track, many of whom also supplied other
venues. Friends and family weren't placated by this statistical
dismissal of the case, and you can't blame them, given that dismissal of fears about these diseases has been a feature of their
investigation.

Other countries have reported clusters as well, but they are
damnably difficult to analyze. For one thing, a cluster doesn't
mean that the people in it have anything in common other
than that they live close together now. The disease takes years
or decades to develop, so even if you find three or four CJD patients living in the same neighborhood, it's quite possible that
if you track backward to when the disease began to incubate
in each of them, they might have been in far-flung places. So
the clustering tells you nothing. In France, one of the other
issues was the difference in average age from city to country.
Older people get CJD; younger people live in the cities, so an

excess of cases in rural areas might represent nothing more significant than age distribution. The bottom line is that, so far at least, a correlation has never been found between a cluster of CJD in humans and an outbreak of prion disease, like BSE, in cows, sheep or pigs.

And sometimes there are clusters that just make you wonder. In 1997, in the medical journal *The Lancet,* scientists at the University of Kentucky reported what seemed to be an anomalous cluster of cases of CJD.[3] They had recorded five patients over three and a half years in western Kentucky (although by the time the article was published they had rounded up another six cases). All had eaten squirrel brains. (In that part of the country, the brains are generally scrambled with eggs and cooked together with vegetables in a stew called a burgoo.)[4] Now, there is no evidence that squirrels carry any sort of misfolded prion, nor was it clear that this was a legitimate, statistically genuine cluster. Clearly more research was needed, but the article did end with a useful piece of advice: "In the meantime caution might be exercised in the ingestion of this arboreal rodent."

BSE (Mad Cow Disease)

And then there's BSE. At first it was just *assumed* that it was the scrapie prion that had somehow vaulted the species barrier—and proliferated—courtesy of meat and bonemeal. What else could it be? It had to be scrapie—it was all over the country and, of course, was similar in many ways to BSE. It was the logical choice: several cows in far-flung parts of the country had come down with the disease almost simultaneously, and scrapie was the only prion disease you'd find lurking in all those places.

But there were some things that didn't sit quite right. One was that in other European countries with scrapie where sheep were recycled to feed cattle, the cattle hadn't had BSE until they got it from the United Kingdom. If it came from scrapie, shouldn't it—couldn't it—have appeared in those other countries independently?

Maybe not. The counterargument was that these other countries just didn't have enough sheep. The United Kingdom had more sheep than any other European country, proportionally more scrapie too, and so a greater opportunity for infection and spread.

How important were changes in rendering and processing in the early 1980s? Was it relevant that meat and bonemeal were being fed to very young calves in the United Kingdom? (Only older cows were fed the stuff in Europe.) The epidemiology suggested that most cows caught BSE when they *were* calves.

Add to that the molecular data that suggested that the two prions were not closely related, and what had been the obvious choice, scrapie, seemed less sure. And no one has yet succeeded in infecting cows by administering scrapie-laden material orally, as would have had to happen for BSE to take off.

Even the logic that scrapie was the only prion distributed widely enough to have launched the epidemic in widely separated locations began to weaken as new and better data pushed further upstream in the epidemic and suggested that there might have been one index case of BSE, one cow that then spread it to all others. In a way, that particular advance just made things more difficult. A single cow could have got some prion from practically anywhere. Remember Jeffrey's nyala? It died of BSE. But you could just as easily argue that maybe some antelope on a game farm got a prion disease, died like the nyala did, was made into meat and bonemeal and *that* was how it started.

This isn't as far-fetched as you might think. There actually was a theory that there was a reservoir of prion disease in African antelopes, some of which, when imported to the United Kingdom, brought the disease with them. The fact that no one in Africa had ever seen a disease like that isn't as convincing as you'd think, because how would you ever know? Necropsy? A dead antelope in the wild in Africa would last about one minute. On the other hand, if there were diseased antelope and lions ate them, then those lions might come down with BSE too—cats are susceptible.

An antelope source was pretty unlikely but not impossible. More reasonable was the suggestion that it could have been a strain of scrapie not yet identified by scientists, or a scrapie prion that began to morph into something new while it cycled silently in English cattle herds in the early 1980s.

But there are other possibilities. One is that it might have all started with BASE, that alternative asymptomatic mad cow disease of older cows. One cow with atypical BSE was found to have a gene that was the counterpart of a human gene involved in genetic CJD. In that one cow at least, what might have looked like a sporadic case of disease was likely genetic.

Add to that the observation that as you pass BASE through the brains of mice, it begins to morph and look more and more like plain old BSE. It's only a lab experiment, but you do have to wonder if, in the early days of BSE, there might have been an unnoticed cow, genetically predisposed to BASE, that died without ever showing symptoms and was processed into meat and bonemeal. But how would we ever know if this happened or not?

The speculation surrounding the origin of BSE is incomplete without the hypothesis of *human*-induced BSE. Drs. Alan and Nancy Colchester suggested in 2005 that the original prions that triggered the BSE epidemic might have come from the Indian subcontinent—and that they were human.[5] Here's their scenario: throughout the 1960s and 1970s, India and Bangladesh were major suppliers of material for the meat and bonemeal that was fed to cattle in the United Kingdom. There was a huge market for this material among British farmers—especially those with dairy cattle. Hundreds of thousands of tons were imported.

The Colchesters argue that some fraction of that imported material would have included human remains. They sketch out a scenario in which incompletely cremated bodies were thrown into the Ganges by devout Hindus, whereupon they were collected by peasants for the export trade in remains just as they'd gather dead animals (note the appearance in Rohinton Mistry's *A Fine Balance* of a bone collector walking the streets).

Would there be enough cases of CJD in India to suspect that at least one would have been gathered and processed? That's hard to say because the official Indian statistics seem low, whereas the authors suspect the real number was much higher. Are enough bodies really thrown into the Ganges without full cremation? The Colchesters say yes (quoting a report that found sixty bodies along a six-mile stretch of the Ganges over two days in April 2004), although some Indian commentators cast doubt on that, arguing that burial and complete cremation are more common. After processing, would there be enough material to cause infection in British cows? As distasteful as it seems to some, it's not out of the question, but it is unlikely that any future evidence will be gathered to evaluate the idea one way or the other. It would be interesting to inoculate some cattle with material from CJD-infected human brain, though, but feeding it to them? Dr. Neil Cashman, well-known Canadian prion researcher, has described that as "the experiment from hell."[6]

Chronic Wasting Disease (CWD)

Although the first recognition of this disease can be pinpointed with more precision than any other (1967, among captive mule deer in Colorado), it is unlikely that it began there. Surveys of wild deer and elk suggest that the places where CWD is most common today are those places where it began decades ago. If that's true, then Fort Collins, the site of the game farm where the disease was first spotted, cannot be the place of origin; it must be farther north, in an area of well-known mule deer migration corridors along rivers with romantic frontier names: the Laramie, Platte and Cache la Poudre. All are packed together at the east side of the Colorado-Wyoming border. But finding out when CWD started is, no surprise, likely going to be impossible. Calculations based on the spread of the disease suggest that one in a hundred deer have to be infected before we actually find that one and diagnose it. Working backward from the number of infected animals today suggests that the first cases must have arisen in the early 1960s or even before, making it even harder than trying to identify the index case in BSE. With BSE, the problem was that many diseases could have been confused with mad cow, and so the first infected animals were simply ignored. In CWD, the situation is exacerbated because the first several animals would likely be killed and eaten by carnivores, or consumed by scavengers, long before a human would encounter them. It is not impossible that the epidemic had more than one starting point, more than one "index" case, although how those would begin at roughly the same time in different places is anybody's guess.

So the "when" of chronic wasting disease is elusive, but no more than the "what happened." The prion is different from many of the strains of scrapie and also from the BSE prion, so where might it have come from? There were sheep living at the same game farm as the first captive animals discovered back in the 1960s, but apparently they tested negative for scrapie. It is the old familiar story when it comes to origins: it might have been a scrapie agent that infected some deer; maybe it was some weird spongiform agent that was always around at undetectable levels but suddenly took off when it infected some deer; possible too is a spontaneous mutation of a prion gene in some unlucky animal.

Searching for any kind of origins is more often frustrating than rewarding and almost always provisional. But it is also fascinating. And ultimately it is important. Could conditions that seem sporadic in fact be infections? Is there some sort of universal prion that hides from us, only to spring these occasional hints that it's there? It seems unlikely, but a theme in this story has been the danger of dismissing the unlikely.

19

Cats but Not Dogs
When Prions Jump
the Species Barrier

The bank vole is an unremarkable mousey animal distributed widely across Europe. But its bland exterior belies its amazing biology. This animal somehow managed to live through the last ice age by finding a secure refugium in the Carpathian Mountains in mid-Europe. As the ice began to retreat, hordes of voles moved out and colonized large areas of the continent.

But that isn't their most amazing survival story. They are one of the species that, to the great surprise of wildlife biologists, have moved back into the exclusion zone around the nuclear reactor that self-destructed at Chernobyl in April 1986. And even though these little guys have the highest body burdens of radioactive cesium *ever measured* in a mammal, they seem to be doing just fine.

Bank voles also hold a place of honor in the prion story: they are unusually welcoming to them. As a result, they are key players in the effort to understand exactly how prion diseases can spread from one species of animal to another. There is something called the species barrier that makes it difficult for one animal's prion disease to infect another. However, the ease with which prion diseases can be established in the bank vole runs counter to that, suggesting that in this animal there's not much of a barrier.

So, for instance, prions of the human Creutzfeldt-Jakob disease multiply just as fast in a bank vole right off the hillside as they do in specialty mice that have been genetically engineered to have human prion proteins. For the purposes of the experiment, the mice *are* human, but even so, they are no more receptive to human prions than the vole. This makes the bank vole an odd animal indeed.

Compare it with the rabbit. The rabbit's legendary promiscuity doesn't apply to prions: it won't incubate *any* of them successfully. The rabbit is immune; the vole is anything but. Between the two live a variety of animals with a range of resistance and acceptance to prion invasion. Being able to explain these differences among animal hosts would give us a much better understanding of how prions work.

There are strange things on the prion side too, and BSE is a nice example. It has an amazing capacity to infect and kill not just cows (it's notable, though, that *all* cows are vulnerable to it, whereas some breeds of sheep are resistant to scrapie), but humans, antelope and many members of the cat family, including house cats (recall the tragic case of Max), lions and cheetahs. Yet, at the same time, domestic dogs, foxes and wolves, which during the BSE epidemic in the United Kingdom ate the same meat-and-bonemeal-infused food as the cats, remained unaffected. Cats but not dogs. Why one and not the other?

It's tempting to think that the answer is to be found in the way the different prion proteins are built, the way they fold and

how they are persuaded to flip from properly folded to misfolded. There's a line of logic that follows: when normal prion
proteins are "persuaded" to misfold, they do so as a result, almost certainly, of some sort of intimate contact between them
and a prion; contact is better if two shapes are complementary,
and shape is in turn dependent on the sequence of amino acids.
So, boiling that down, the sequences of both prion proteins
(properly folded) and prions (misfolded) are the key to understanding why infection is sometimes easy and sometimes not.

Why else would a bank vole be susceptible to so many different prions and the rabbit to none, and why else would BSE
be so omnipotent? Take the rabbit and the vole: both, like all
mammals, have uncountable normal prion proteins in their
brains. But those proteins are not identical—the two animals
have been evolving separately for something like sixty million
years, and separate runs of mutations in each have altered what
were, if you go back far enough in time, identical sequences.
As amino acid sequences change with time, so too must the
shape of their proteins.

There should be something about the shape of the rabbit
prion protein that makes it inhospitable to prions, and in turn,
something about the vole's prion protein that makes it vulnerable, and there are indeed a couple of amino acids in the
vole protein that appear to be all-important, those at positions
154 and 169, out of a total chain length of well over 200. Voles
have different amino acids at those two positions than, say, lab
mice do. These must be crucial places that somehow, through
the shaping of the protein, determine prion susceptibility.

But what would make those two places exceptionally important? That is one tough question. If it were known exactly how the prion protein misfolds, how the chain twists and turns and reconfigures itself, it would be much easier to examine what roles amino acids 159 and 164 play. But no one has ever seen an individual misfolded prion, and it might be that there really is no such thing, in that they always associate with others, in clusters and fibers that clump together so aggressively they cannot be disaggregated.

Nor is it understood exactly *how* the normal protein becomes converted. Does it come into direct contact—a fatal embrace—with a single misfolded version? Or, more likely, is it coerced into joining a growing crystal-like array of prions, clicking into place like sucrose molecules onto a crystal of sugar? In that scenario, such crystal-like growths (the fibrils that we have already met) would grow, then break into pieces, whereupon each piece would grow and fragment again, all the while increasing the number of sites for misfolding new proteins. But again, no one knows for sure.

This scenario can make some sense of the species barrier. The misfolding event would require some sort of coming together, and the closer the invading prions were in sequence and shape to the host prion proteins, the closer the fit might be and the more efficient the conversion. For instance, scrapie prions are exactly the same amino acid sequence as the prion proteins in a sheep's brain, so the disease process is smooth and efficient. Put scrapie prions in a different kind of brain and things wouldn't work so well.

Of course, if this were completely correct, if absolute synchrony in sequence were necessary, there would be no cross-species infection at all, but as we know, there is: BSE in humans, kuru in chimps, CJD in voles. In each of these cases, neither the time frame nor the sites of greatest damage in the brain nor the symptoms are exactly the same as in the normal host, yet disease takes hold. And sometimes that process is puzzling: when hamsters are inoculated with brain tissue from mink suffering from transmissible mink encephalopathy, two completely different patterns of disease result: drowsy and hyper. (Curiously, in the 1960s, it was recognized that injecting scrapie into goats produced drowsy and scratchy versions.) The prions being injected were all the same, but the genetics of the hamsters weren't uniform: they had slightly different prion proteins, which, when forced into misfolding, took up slightly different shapes, and those shapes caused different symptoms.

Note that in this case, a single kind of prion infecting slightly different animals yielded slightly different prions. They were dubbed "strains," the word being adapted from the study of viruses. Unlike viruses, of course, prions have no genes to mutate and create different strains, so how does that happen in their world? It can't be differences in amino acid sequence because those don't change. What else can happen to a protein to make it behave differently? A shape change. The existence of multiple strains (there are something like twenty different strains of scrapie) implies that there must be twenty different ways of folding, refolding and *mis*folding the sheep prion protein. This is where the science departs so radically from what was thought decades ago.[1]

If you had tried to persuade a protein scientist of that in the 1960s, you would have been dismissed as ignorant or crazy. It was widely believed then that proteins adopted the one shape that best accounted for the chemical pushes and pulls created by its string of amino acids, and that there was really only one way of doing that to stabilize the molecule best. The new idea: proteins may indeed have just one best shape, but a population of them is apparently heterogeneous, with many different shapes clinging to existence. They're not the most stable, but they're there.

Recent experiments have shown that when a population of prions, this mix of slightly different shapes, is challenged by, say, a new environment, sometimes a minority shape steps to the fore, flourishes and becomes predominant. This happens, for instance, when prions from brain tissue are thrust into cell culture. It's a different medium and, while it takes time, new variants of prions appear that are better adapted to cell culture. Put them back in brain tissue and the population shifts back to its original composition.

If this sounds like natural selection and evolution to you, you're not the only one. Some scientists argue that this is exactly what's happening: prions evolve. Start attacking them with a drug and there will definitely be an immediate die-off, but in time the population struggles back, led by variants that were unimportant before.

So if they can evolve, are they alive? Most scientists would say no because they, like viruses, cannot reproduce outside their host cells or tissues. But they are awfully like living things.

This research makes more concrete the idea that it's quite likely that a population of prion proteins consists of one majority shape (the best one) and many others, some close to the best shape, some not. Maybe one molecule in every ten thousand is, if not unfolded (and therefore susceptible to *mis*folding), at least in some different state than the majority.

But it's more dramatic than that: we have solid evidence that even if one's brain is completely healthy and normal, there are clusters of misfolded prions in it. "Silent prions lying dormant in normal human brains" was the way the researchers put it in their report of their findings. They also wrote that "the difference between prion-infected and uninfected brains seems to be quantitative rather than qualitative." These apparently harmless prions might be "awaiting a change in state of the host or transmittal to a new, more susceptible host."[2] Maybe it's true that our brains are always in an uneasy balance, harboring silent prions and most of the time minimizing their presence by clearing them away as fast as they form. But invading prions, some sort of biochemical event, a mutation—any of these could tip the balance.

It's also true that these minority shapes are changing from moment to moment, sometimes existing in very unlikely, awkward, unsustainable versions, sometimes adopting very stable conformations. It is in this suite of shapes that the vulnerability to misfolding lies.

A misfolded prion is not its natural shape, nor is it a shape that the normal protein is likely to adopt spontaneously. Misfolding requires some sort of biochemical coercion, exerted by either another misfolded prion or a growing crystal of them, perhaps to get over an energy hump. Presumably, that transformation is easier when the shape of the normal protein is similar to the misfolded one, or at least when the transition from one to the other is relatively easy. If the two are absolutely identical proteins, each amino acid matching perfectly, then the conversion process should be rapid and efficient because their shapes will be nearly identical. But when the infecting prion differs at several positions along the chain from the normal version, conversion will be much more difficult and slower.

This shape-changing explains why hamsters can generate two different strains of mink disease: there are two slightly different types of native hamster prions, each of which adopts a slightly different shape and thus is misfolded slightly differently by the same invading prion.

So back to the bank vole. There's something about its normal prion proteins that makes them easy to misfold, whether it's the native shape, the sequence of amino acids—something. Whatever that something is, it's missing in rabbit and dogs, whose prion proteins misfold with great difficulty or not at all. We humans are somewhere in the middle: vulnerable to BSE prions but apparently resistant to scrapie. In fact, BSE is the only nonhuman prion disease that has been shown, thus far, to infect humans. Our vulnerability to the prions of chronic wasting disease is unknown, although experiments so far suggest we are resistant to the disease—but more on that later.

Even the species barrier is mysterious. As mentioned earlier, when a prion disease spreads from one species to another, it usually takes much, much longer to establish disease in the new host upon inoculation than it does in the species from which it came. But with repeated attempts at transmitting the disease, that species barrier breaks down, and the incubation

period shrinks. (Although some species remain forever resistant, such as dogs with BSE.) Apparently, once misfolding has
occurred once, it becomes easier. Furthermore, once the host
proteins have been misfolded into replicas of the invaders,
they'll maintain that shape when inoculated into other species.

Here's the perfect example. Sporadic CJD, the classical
human prion disease, can be transmitted to mice only if those
mice are genetically engineered to carry the human gene for
the prion protein, not their own. Normal "wild-type" mice are
virtually resistant. But variant CJD, the one brought on by eating BSE-contaminated meat, is the absolute reverse: it infects
wild-type mice easily, but mice engineered to be "human" only
with great difficulty. What gives? Remember that variant CJD
is the human version of mad cow disease. What must be happening here is that when humans get this disease, their prion
proteins are forced into a shape that remembers or reflects the
mad cow prion, and that is a shape different from that of the
prion of sporadic CJD. It is true that mad cow prions can infect a very wide variety of animals. If you performed the mas-
sive experiment of inoculating them all, then collected all the
prions that result—from all these animals—and injected them
all into mice, you get exactly the same strain, same areas of
damage in the brain, same incubation period. It's as if the BSE
prion passes largely unscathed through all these other animals
even though their prion proteins are all different. Maybe, just
maybe, this is a prion conformation that is unusually stable.
After all, in the 1980s and 1990s, it survived the trial by fire of
passing through rendering plants and their cooking processes.

Let me close this chapter with three intriguing puzzles that suggest we're a long way from truly understanding this kind of infection. Remember that variant CJD, the form caused by eating BSE-infected meat, is quite different from sporadic CJD. Its prion is much more like the BSE prion—that was the smoking gun suggesting that the one came from the other—and the symptoms are very different.[3] Victims are usually younger, the disease lasts longer, the destruction of the brain has features never seen in sporadic CJD and—this is the big thing—the prions themselves are different. But experiments have shown that if mice are inoculated with BSE prions, some of the new prions that appear as the disease takes hold are like variant CJD, as you'd expect, but some of them are more like sporadic CJD.[4] These are weird and unsettling results. With the exception of some mavericks, no one thinks that sporadic CJD, that one-in-a-million worldwide prion disease, has anything to do with BSE or, more to the point, that it comes from eating infected beef. After all, CJD long predates the BSE epidemic, and it occurs in countries where there is no BSE. Yet, these results show that it's conceivable that there could be cases, say, in England, of what looks like sporadic CJD that actually have come from BSE. Indeed, there are data that suggest that the rates of sporadic CJD in England have risen since 1970, although these higher numbers are usually explained by the fact that the disease is being monitored more closely, not that some cases of sporadic CJD have actually been caused by eating BSE-infected beef.

Another set of experiments in Scotland showed that certain mice, inoculated with either a strain of human prion disease or scrapie, came down with unmistakable prion disease, which could then be easily transmitted to other mice. All of that is completely straightforward. The problem was that the amounts of detectable prions were vanishingly small, virtually undetectable by some of the standard methods of prion chemistry.[5] The authors of the paper describing this experiment made the understated point that it is not a good thing to have prion diseases that would be overlooked by the usual methods of detection.

And finally, on a rare positive note, one recent and very dramatic experiment showed that disease caused by the misfolding of prions can be not only halted in its tracks but actually reversed. This was one of those rare experiments where the design was almost as cool as the results. It built on the observation years ago that mice lacking the ability to synthesize their own prion protein are completely resistant to infection by prions. This observation became one of the key pieces of evidence for the idea that these diseases depend on the conversion of host prion proteins by the invaders. In this more recent version, researchers engineered mice so they had a genetic switch that could turn off the production of the prion protein in their brains.[6]

Both these engineered mice and normal, control mice were infected with prions. As expected, all began to show signs of disease: their behavior changed—they developed memory problems and their tendency to burrow declined—and at the same time their brains began to show the typical signs of prion disease. Then the genetic switch was activated in the experimental mice, and they responded by shutting down their prion protein production. In turn, the levels of prions in their brains began to drop because they had no more targets to misfold. But more important, there was also a striking change in the health of the mice. Not only did the progression of the disease stop (while continuing unabated in the controls) but it went into reverse. They regained the ability to recognize novel objects, they returned to normal burrowing behavior and their brains regained a normal appearance. In effect, they were cured of their disease.

Before leaping to the conclusion that this opens up the possibility of treating such diseases in humans, here are a couple of cautions. First, we don't know how to turn off prion protein production in humans, and even if we did, we don't know if that would be a good thing. You have to assume that they do something useful, although it's not yet clear exactly what it is. Second, it's not yet known whether more advanced disease, at the stage it would likely be diagnosed in humans, could have the clock turned back. But at least it's a start, a piece of knowledge, an idea of how we might think about treating these invariably fatal diseases.

And that foothold, as tentative as it is, points directly to one of the most intriguing directions that prion studies are now taking: to other diseases that, while not infectious, do involve misfolded proteins and are much more prevalent, much more crucial to health care in the future. They include amyotrophic lateral sclerosis (Lou Gehrig's disease), Parkinson's disease, chronic traumatic encephalopathy and, the most prevalent of all, Alzheimer's disease.

20

Alzheimer's Disease
Plaques and Tangles but
So Far No Prions

In 1982, when Stan Prusiner published his famous paper in the journal *Science* introducing his "prions" to the world, he claimed that future research on them might hold the key to a set of other human diseases, including ones much more common and with much greater societal impact than CJD or kuru. It was a substantial list: "Alzheimer's senile dementia, multiple sclerosis, Parkinson's disease, amyotrophic lateral sclerosis, diabetes mellitus, rheumatoid arthritis, and lupus erythematosus, as well as a variety of neoplastic disorders."[1]

It was a bold statement. Some would say it was dramatically oversold. However, either it was just dumb luck (well, relatively dumb) or Prusiner was even smarter than those faithful to him would argue, because the vast majority of scientific evidence that would give support to that prediction has surfaced only in the past few years. In the end, Stan might have been right—again. There are at least four diseases that bear enough resemblance to the prion illnesses to suggest that they might benefit from the knowledge of how prions disrupt and kill. These include ALS (amyotrophic lateral sclerosis, or Lou

Gehrig's disease), chronic traumatic encephalopathy and Parkinson's disease, and all deserve a closer look. But for obvious public health reasons, the disease that leads the list is Alzheimer's disease.

Six million North Americans have Alzheimer's, but by 2050 that number could be eighteen million. One out of every two people over eighty-five has it. Obviously, any leads whatsoever that would get us a little closer to an effective treatment would be huge. So Prusiner knew that a list of diseases with Alzheimer's at the top would attract attention, but it turned out he wasn't just blowing smoke.

There are tantalizing similarities between Alzheimer's and the human prion diseases. Both attack the brain, whereupon brain cells become deformed and die. In Alzheimer's, dark spots appear in the brain either between cells or inside them; the between-cell spots are plaques, clumps of misfolded proteins. They look pretty much like those found in CJD, vCJD and kuru. The resemblance seems to extend even to genetics: people who have certain amino acid combinations at that famous site 129 in the human prion protein seem to be more susceptible to early-onset Alzheimer's disease and at the same time are prone to short and vicious versions of CJD. You might be tempted into thinking that if a particular mutation in the prion gene plays a role in Alzheimer's, there's a connection worth following up. And there might be. On the other hand, in one way, Alzheimer's and CJD or kuru couldn't be more different: you can't catch Alzheimer's.

Or can you? There are people who think that Alzheimer's disease *is* infectious. Prominent among them is Dr. Murray

Waldman, whose book *Dying for a Hamburger* (written with Marjorie Lamb) argues that Alzheimer's has become an epidemic disease only in the past century and that its dramatic increase in numbers parallels changes in the way beef is produced and consumed.[2] BSE all over again. Waldman points out the similarities between Alzheimer's and the prion diseases, but much more emphatically, claiming the similarities "are so strong that it appears obvious they must be connected."[3]

Not just connected: Waldman argues that both *are* prion diseases and that we come down with either one by eating meat contaminated with prions. If this were true, it would mean that the tens of millions of cases of Alzheimer's would have been the horrific result of eating infected meat. Waldman arrives at this conclusion by taking the well-attested example of variant CJD, which is indeed caused by consuming BSE prions, and extends that model both to sporadic CJD (where there's no evidence of transmission or infection) and to Alzheimer's disease (where there's no evidence of transmission, infection or even prions).

It's true that people incubating CJD can transmit the disease, as happened with growth hormone, and also that people with vCJD can transmit the disease, as happened with blood transfusions. But where is the evidence that people "incubating" Alzheimer's have transmitted their disease to others? Waldman argues that there are only two things that have been identified as causing CJD: eating meat (presumably he's talking about vCJD) and the cases of friendly fire: infected growth hormone, dura mater and surgical instruments. He then argues that because only 5 percent of all cases of CJD have been attributed to those friendly fire accidents, it's reasonable to assume that all the rest are due to eating meat. But there are no data to support the idea that when we're eating meat we're consuming prions that cause Alzheimer's disease.

At the same time, Waldman makes a claim for some interesting curiosities: Why was Alzheimer's virtually unknown before the nineteenth century? Why has it spread throughout the world—like an epidemic—if it isn't infectious? And why, in countries where the rates of CJD are low, are the rates of Alzheimer's low as well? Obviously, there could be all kinds of mitigating factors operating here, from inadequate sampling and/or diagnosis to lower life expectancies, as in Africa, where death usually would come before either of these diseases appeared.

But aside from the assertions of Waldman and a handful of others, there is nothing linking Alzheimer's disease to CJD other than the superficial similarity, and no evidence that Alzheimer's disease is infectious. So was Prusiner unjustified in claiming that light would be shed on Alzheimer's and other diseases by prion research? Not quite: there are, at the molecular level, intriguing connections whose significance isn't yet exactly clear. Understanding those similarities is proving to be a difficult challenge, however, because Alzheimer's disease is still mysterious and incredibly complex. There are two features of the disease that seem to set it apart from prion diseases like CJD. One is that the association of plaques in the brain and dementia is not as tight as it might at first have seemed; the other is that, while Alzheimer's presents as a disease of the elderly, there is evidence to suggest that its beginnings are to be found decades earlier.

For instance, most autopsies of patients who died with Alzheimer's disease will reveal a brain riddled with two kinds of debris: plaques and tangles. The tangles are twisted pairs of fibers made of a protein called tau that accumulate inside neurons; the plaques, dark spots, accumulations of misfolded proteins not unlike those seen in kuru, laid down among the neurons. The Alzheimer's brain is full of plaques and tangles; the unaffected brain is not. But doubt has been cast on what this means by a long-term study of nuns in the United States.

In the early 1990s, 678 nuns in the American convent School Sisters of Notre Dame were enrolled in this study: they took regular tests to evaluate their memory and mental acuity, and also pledged to allow their brains to be autopsied. The goal of the study was to track how the mind changes with age; the huge advantage of doing this with the School Sisters is that they are an unusually homogeneous group, environmentally if not genetically, eating the same food, keeping the same schedules and taking part in similar activities, day in and day out. Many of the confounding variables that would render most studies like this unbearably complex are thus no longer on the table.

The School Sisters have turned out to be both fascinating and paradoxical. Sister Mary is a good example.[4] She died at the age of 101. A few months earlier, she had taken her last set of tests, including the Mini-Mental State Examination (MMSE), the Boston Naming Test and a verbal fluency test. She did amazingly well, scoring twenty-seven on the MMSE, a score that ranks her in the normal range (scores ranging from twenty-six down represent increasing degrees of dementia).

Considering she had less formal education than 85 percent of the School Sisters (and education does influence the score no matter how old you are when you take it) and that she took this test only months before she died, Sister Mary was obviously a rare individual. In fact, her predicted score, based on her age and her relative lack of formal education, was not twenty-seven but four! This isn't to say she was as fluent as she once was or had the memory she once had, but she was one of those rare individuals, a centenarian with no signs of Alzheimer's disease.

Although she was mentally intact at 101, the autopsy of her brain revealed a heavy load of both tangles and plaques, more than enough to have been incapacitating. There is such a well-established correlation between the symptoms of dementia—confusion, loss of memory, disorientation, poor judgment, changes in personality—and those plaques and tangles in the brain, if you had seen her brain and the telltale deposits, you would have been confident that she had been demented. The Alzheimer's brain is usually full of both; conversely, those who die undemented are expected to have very few. Sister Mary was different. She was one of the most paradoxical examples, but by no means the only one who resisted the onset of dementia despite having the telltale deposits in the brain.

How to explain this? Protection might be afforded by something called brain reserve, a two-part hedge against the descent into dementia. One part is physiology: a well-developed brain, full of neurons and the innumerable links they make with each other. Brains like that are the result of good nutrition, social environment, education and genes. The other side of brain reserve is reaction. An example would be switching to different neural circuits to compensate for those damaged by, in the case of dementia, blood clots, plaques or tangles.[5] Brain reserve is an appealing idea: it suggests that *some* brains can withstand the impact of accumulating damage, at least in the case of Alzheimer's. Is there a parallel to human prion diseases? Not really: there is the evidence that having the right genes that build in the best amino acids at site 129 in the prion protein can delay death for years or even decades, but that likely has more to do with the shape similarity between invading prions and host proteins, and not the prion equivalent of brain reserve.

This disconnect between plaques in the brain and symptoms of Alzheimer's is one important piece of evidence supplied by the "Nun Study"; the other is the shocking idea that the disease has its roots in a person's early twenties. Beginning in 1930, young women entering the School Sisters of Notre Dame were required to write an autobiographical sketch, including references to parents, education, interesting or crucial life events and an explanation of why they chose to enter the convent. These brief essays have turned out to be a valuable resource, a record of a mental starting point for each nun in the study. And here's the remarkable thing: analyses of these essays, written when the nuns were twenty-one or twenty-two years old, can be used to predict whether or not those women will, when they're in their eighties and nineties, become demented. It sounds unbelievable that a short exercise in self-expression can predict, with something like 90 percent accuracy, mental states seventy years later, but it can and it does.

These essays have been evaluated for two things: idea density (the amount of information in a sentence), and grammatical complexity (the number of embedded clauses in those sentences). The two tap into different brain mechanisms. Idea density is a subtle feature of expression that is measured as the number of ideas expressed every ten words.[6] Grammatical complexity stresses what's called working memory. If you've ever had the experience of starting a complicated sentence, then realizing partway through that you've lost your train of thought, that's an example of your working memory not being quite up to the task of grammatical complexity you set for yourself. Working memory is what you're using when you try to remember a phone number you've just heard. As it declines with age, so does the grammatical complexity of written work.

This isn't the only evidence that Alzheimer's develops throughout life. One remarkable study in Scotland showed that children's scores in a 1932 intelligence test—when they were eleven years old—correlated to their likelihood of getting Alzheimer's disease late in life. The better the score, the lower the risk. Other studies have shown that plaques appear in our brains by the time we're in our forties, and the accumulation of neurofibrillary tangles begins even earlier, in our twenties.

All this goes to show that Alzheimer's disease is not a straightforward one-to-one case of brain pathology correlating to symptoms. You may be heading for the disease without any physiological indicators at all, but you can also have those disease markers with no apparent disease. Is Alzheimer's in this sense anything like the prion diseases? Superficially, the best comparison is with Creutzfeldt-Jakob disease, but let's take a closer look.

Both are largely, but not exclusively, diseases of old age. Both involve dementia and loss of memory, and both result in the accumulation of plaques in the brain. They are also somewhat similar in that the prolonged incubation period in the prion diseases could be likened, at least superficially, to the decades-long development of Alzheimer's disease. And genes play a role in both. In Alzheimer's there are genes that are known to increase the risk of getting the disease. Genes are more definitive in the familial prion diseases, like familial CJD and fatal familial insomnia: if you have the gene, you get the disease and you die.

But there are also significant differences between Alzheimer's and CJD. The numbers, for example. In North America there are over 300 cases of sporadic CJD a year, but more than *360,000* new cases of Alzheimer's disease. And while that ratio might be in the neighborhood of a thousand to one right now, the rate of Alzheimer's is rising rapidly, whereas the rate of CJD is not. Even if you take seriously claims that a significant number of Alzheimer's deaths are in fact cases of CJD (one study claimed that number was 13 percent), the difference is still huge and growing. Incidentally, when you take into account that the average case of Alzheimer's lasts nine years, as opposed to CJD's six months, it's obvious which is the bigger burden on health care systems.

Second, there is no evidence that the eventual appearance of symptoms in a prion disease like CJD can be prevented by mental activity or education. It can be delayed by genetics, as in the locus 129 story in kuru and vCJD, but that's a very different thing. The major difference, and at the moment the showstopper, is that the prion diseases can be transmitted. Kuru, CJD and vCJD and their animal counterparts can all be inoculated into lab animals and cause disease. And all three of these human prion diseases have been inadvertently transmitted person to person.

There is no evidence of Alzheimer's disease being caught. Medical personnel who deal with Alzheimer's patients and those patients' families don't appear to have a heightened risk (although with 50 percent of all people over age eight-five getting the disease, it is difficult to tell), but neither do people in close contact with patients with prion diseases.

So the diseases have some similarities in symptoms and behavior, but the numbers are very different, and transmission is possible in one but seems not to happen in the other. What about the destruction of the brain? It's there that you see some curious connections.

Kuru, CJD, variant CJD and Alzheimer's disease all feature deposits of accumulations of misfolded proteins called plaques, although in none of these diseases is it clear exactly what role the plaques play. In the prion diseases, most researchers lean toward the belief that the plaques are not active sites of infection, but rather dead-end accumulations of misfolded proteins. Evidence from several directions suggests that the actual infectious agents are small aggregates of individual prions, fewer than twenty.

In the lab, prions begin clustering together, eventually forming short rods that continue to grow in length. Once they get to a certain length, the rods aren't infectious, but if they are broken up into smaller pieces, those pieces are. It isn't likely that the misfolded proteins in plaques are actively recruiting new misfolded proteins. It *is* likely that the damage to neurons—brain cells—is due to the short rods. So when it comes to prion diseases, seeing plaques in brain tissue means the damage is already done. It may be ongoing, but plaques are preceded by the much more dangerous and much smaller aggregates.

The plaques seen in Alzheimer's are different. They are agglomerations of misfolded proteins, true, but not prion proteins. They are composed of something called beta-amyloid, and their role in the disease is controversial. As mentioned earlier, they have always been viewed, along with the tangles inside brain cells, as the most solid evidence of Alzheimer's, but are they cause or result? It's true the amyloid might interfere with neurons, maybe even kill them. On the other hand, lumps of amyloid could just be a by-product, an accumulation of garbage left behind as the disease spreads. But dangerous garbage: most researchers would buy the idea that plaques play some role in creating the chaos and damage that typify Alzheimer's. And it's garbage that sets some sort of speed record. Recently, researchers created tiny windows into a mouse brain and, using high-resolution microscopes, were able to watch as plaques virtually popped out of nowhere. One day, nothing; the next, plaque. Then, after twenty-four hours, the plaques stopped growing.

Do these Alzheimer plaques share anything significant with the plaques of, say, kuru or CJD? In some special settings, they can be made to act like prions: one experiment used two strains of mice engineered to carry human genes that predispose the animals to develop plaques when they age. The kind of plaque is different in the two strains. When these mice were inoculated with material taken from the brains of people who had died of Alzheimer's, or from the other strain of mice, plaques began to appear in large numbers. The appearance of the plaques differed depending on which strain of mouse was the donor and which was the recipient. That is the same sort of picture you'd get with prions. Also, the plaques spread from their initial locations throughout the hemisphere of the brain that had been inoculated. It looked like transmission of the disease, but obviously this was a very special case; because the experimental mice are engineered to develop Alzheimer's later in life anyway, it could be that the injection simply accelerated—by maybe a few months—what was inevitable anyway.[7]

But as I was writing this chapter, another and much more startling experiment was published. It was much like the one I just described, except that these mice carried human genes that did *not* predispose them to develop their own plaques. In other words, these mice would normally live a long and healthy life and never develop plaques. Yet, when inoculated with Alzheimer's material from a ninety-year-old woman who had died from the disease, these mice had fully developed Alzheimer's plaques in their brains after about a year and a half. It also appeared that early-stage plaques had started to appear much sooner than that, and plaques were found in parts of the brain distant from the inoculation site. By contrast, mice inoculated with brain material containing no plaques were unaffected.

This experiment made the headlines, because it seemed to suggest that Alzheimer's disease can be infectious, at least in circumstances where plaques are introduced directly into the brain. How would that happen? The story of CJD and the contaminated electrodes comes to mind immediately, but there's no evidence that electrodes could pick up plaques—as opposed to prions—from a brain and there are techniques for sterilizing surgical instruments that have been put in place since the cases of CJD transmitted by electrodes (so the likelihood should be small); blood transfusions might offer another route (but then brain plaques would have to be circulating in the blood). This is really just speculation.

It's also important to point out that plaques are only half the story; there are also the diagnostic tangles found inside brain cells, twisted fibers of misfolded proteins, different from the plaques and apparently capable of spreading throughout the brain. The usual function of the tangle protein, tau, is to create and organize the cell's skeleton, the microtubule network, responsible for maintaining the shape and structural integrity of the elongate neurons. They too have a role.

Here's what can be said at the moment: at the molecular level there appear to be significant similarities between prion diseases and Alzheimer's disease. There are misfolded proteins in both, and under the right circumstances, those proteins (at least the plaques in Alzheimer's disease) appear to be able to spread, in both number and location. So it might be possible that therapies based on these similarities could be developed. Even though at the moment there are no effective treatments for either, any possible connection to Alzheimer's should spur research into the prion diseases, and no one knows where the ultimate answers will come from.

There are these similarities between Alzheimer's and prion diseases, but they don't add up to saying that Alzheimer's disease is an undiscovered prion disease. Nor is Alzheimer's disease a prion disease spread by both the consumption of meat and the use of the huge variety of products that are derived from meat. If it's a prion disease, let's see some full-on prions!

21

Parkinson's Disease
Looking More and More Like a Prion Disease

Parkinson's disease might be, at first glance, more straightforward than Alzheimer's. One small area of the brain, the substantia nigra, begins to die, and the supply of the neurotransmitter it produces, dopamine, shrinks. The depletion of dopamine is responsible for the characteristic symptoms of the disease: tremor, rigidity, loss of balance and slowing movements. But again, the disease spreads to other parts of the brain, and ultimately other symptoms, like dementia, begin to appear.

There is a remedy, at least in the short term: replace the dopamine. Once this chemical is back up to normal levels, the symptoms ease. But achieving the right levels of dopamine is tricky: too much and patients can develop symptoms of psychosis; too little and the patient is still suffering. And over the long term, as natural dopamine levels continue to fall, replacement becomes trickier, especially because the more dopamine is supplied, the less the brain makes.

In this simple view of the disease, dopamine is front and center, and that's as it should be. It's important to note, however, that this disease is in its own way as complicated as Alzheimer's. For instance, it's long been known that long-term exposure to certain herbicides and pesticides, especially rotenone, a common insecticide, and Paraquat, which was controversially used to kill marijuana crops in Mexico in the late 1970s, raises the risk of getting Parkinson's. Farmworkers especially are at higher risk for the disease. By contrast, substantial research shows that cigarette smoking and drinking coffee are protective, likely because of the nicotine and caffeine. Add to this growing evidence of various genetic susceptibilities to Parkinson's and it's easy to accept that this is a disease that must be approached from a molecular perspective: it is only logical that all of these chemicals exert their diverse effects, pro and con, on neurons, that the genetic predispositions have their roots in proteins and that it is in those same neurons that the manifestations of the disease can be found.[1]

Abnormal protein deposits are found inside the brain cells affected by Parkinson's disease. They are (this is all too familiar) accumulations of misfolded proteins but, once again, unique to the disease. The protein that misfolds in Parkinson's disease is called alpha-synuclein; innumerable misfolded copies of alpha-synuclein clumped together are called Lewy bodies. But it isn't enough simply to discover that there are protein aggregates like this: they must somehow be integrated into a sensible story of how Parkinson's begins and spreads. And here is where the story takes on scarcely believable twists and turns.

In the late 1980s, Swedish clinicians began to try transplanting brain tissue from aborted fetuses into the brains of Parkinson's patients. Their hope was that if these transplants took, they would supply the missing dopamine. The results were mixed. In many cases, the patients developed severe involuntary movements (which only recently were explained by the presence of neurons in the transplants that exaggerated the release of dopamine). But in some instances, the impact was extraordinary: transplants were given to two Americans who had acquired severe Parkinson's by injecting a badly prepared street drug contaminated by a toxic chemical by-product.[2] That contaminant had virtually erased the part of the brain that is involved in the disease, and no amount of dopamine replacement by the usual methods was going to help. Their recovery was stunning: one of them, a forty-three-year-old male who had been virtually immobile, was able to ride a bicycle within six months.

There were other successes too, although none quite as spectacular, and several people lived ten or more years after their transplants in reasonable comfort. It was at the autopsy of two of these patients, one of whom had survived sixteen years after transplantation, the other thirteen years, that a hint of prion-like activity was revealed.[3] The autopsies showed that Lewy bodies (those accumulations of alpha-synuclein characteristic of the disease) were present not just in the patient's own brain cells but in the transplanted ones as well. This was shocking because those cells were only about as old as the transplants themselves, and Lewy bodies are rarely if ever present in the brains of thirteen- or sixteen-year-olds. It appeared that the disease had spread from the neighboring diseased tissue into the transplants. The immediate reaction was, of course, dismay: this cast a shadow over the promise of transplants for Parkinson's, since if the disease spreads into the new tissue, that tissue will inevitably fail as well.

Adding to the dismay was the realization that here was another indicator of misfolded proteins having the ability to move from one cell to another, possibly explaining in yet another neurodegenerative disease why and how the disease spreads. Further studies in the lab showed that it indeed was true that if you cultured cells laden with Lewy bodies adjacent to untouched cells, in time you'd find Lewy bodies in all of them.

That was not the end of the story. Recall that in mad cow disease, the invading prions must somehow travel a tortuous route from the stomach of the cow (which ate the infected meat and bonemeal) gradually but inexorably to the brain, a journey with enough twists and turns that it's not surprising the incubation period of the disease is so long.

At first glance, it's much the same in Parkinson's disease. Accumulations of Lewy bodies have been seen first in the nerves around the gut and in the spinal cord, long before there's any sign of them in the brain. But they do move up the spinal cord, spreading first through the lower parts of the brain closest to the cord. The vulnerable substantia nigra is one of the prime targets there. With time, Lewy bodies can be found in the highest reaches of the cerebral cortex. When accumulations reach critical levels, disease sets in.

The idea that the Lewy bodies could migrate all the way from the gut nervous system to the brain might make sense of that long-standing observation that environmental factors are linked to the disease. A prominent example is the observation that farmworkers, especially those who use herbicides and pesticides, are uniquely at risk for Parkinson's. It might be that those chemicals need only to get into the body somewhere to trigger the disease, with spread of the disease to the brain following after. Tracking down what's going on there has been difficult, but a study of mice being fed small, regular doses of the pesticide rotenone suddenly makes sense of it.[4] It took nearly two months or more, but the mice started to lose their balance on a rotating wheel (the mouse equivalent of log-rolling), and when their brains were examined, there was alpha-synuclein— the stuff of Lewy bodies—everywhere: in the nerves of the gut, in the spinal cord and, after three months, in the substantia nigra itself. The longer the mice consumed the chemical, the further the aggregates reached, but at no time, even with incredibly sophisticated and sensitive instruments, could any pesticide be detected in the blood or brain, nor did there seem to be any damage caused by the drug itself. The damage was being caused by the misfolded proteins spreading through the nervous system, and they in turn were somehow the result of ingesting a pesticide. A startling experiment.

Of course, with Parkinson's, as with Alzheimer's and amyotrophic lateral sclerosis (Lou Gehrig's disease), there is still no evidence of infection: no one *catches* this disease. But there is so much that is reminiscent of the prion diseases that naturally researchers wonder if it might be possible somehow to stabilize proteins in the cell, make them resistant to misfolding and so halt or even reverse the disease. That is a long way off, but there are more similarities to prion diseases than differences being discovered every day.

22

Lou Gehrig's Disease
The Emerging Picture of
a Prion-Like Process in ALS

Lou Gehrig's is another, almost always fatal, neurodegenerative condition that has its prion connection. The famous Yankee first baseman is said to have succumbed to this disease, the technical name of which is amyotrophic lateral sclerosis (ALS). Recent research has raised questions about that, however.[1] A group of American researchers showed that not only can repeated blows to the head create molecular mayhem in the brain, causing a disease called chronic traumatic encephalopathy, but that in some cases the damage extends to the spinal cord, and such a patient could be diagnosed as having ALS. It was already known that professional athletes, especially soccer players in Italy and football players in North America, seemed to experience unusually high rates of ALS, but this study made the direct link between damaged tissue and the disease.

Lou Gehrig played football when he was young and certainly was hit on the head by pitches in his lengthy pro career. His extraordinary 2,130 consecutive game stretch with the Yankees was marked by at least one instance (but probably more) when he was nearly knocked unconscious by a pitch but stayed in the game. Gehrig was cremated, so we will never know exactly what caused his death, but there is a possibility that it wasn't the typical ALS (although the aforementioned American researchers point out that *they* never suggested this).

Aside from calling into question the linking of Gehrig's name to ALS, the American study suggests that blows to the head may occasionally precipitate a disease that could be indistinguishable from classical ALS. Both are diseases of the motor neurons, those cells that carry messages from the brain to the muscles, commanding them to contract. As these cells gradually degenerate and die, the symptoms progress from muscle weakness, clumsiness and cramping. It can start in the limbs and erode the ability to move, ending in paralysis or, more centrally, affecting speech, swallowing and eventually breathing. Patients with ALS, the remarkable Stephen Hawking notwithstanding, usually die within five years.

Here too, as in Alzheimer's, there are tantalizing molecular resemblances to the prion diseases. At the heart of ALS is yet another protein with a different name (this one is superoxide dismutase) that normally helps to protect cells from destruction by free radicals. Free radicals are notorious for being highly reactive and destructive, and a prime example is the free radical called superoxide. We put ourselves at risk because our own cells produce lots of it. How much risk? Mice with mutations that prevent them from getting rid of superoxide are prone to cancers and cataracts, lose muscle mass or just die young of oxidative stress.

The job of superoxide dismutase (SOD) is to disarm these little sticks of cellular dynamite by turning them into hydrogen peroxide and oxygen. It is very good at that job too, as long as it maintains the right shape. The importance of shape to this particular protein is underlined by the fact that it is extremely resistant to treatments that normally disrupt proteins.

The problem starts when *mutant* versions of SOD are created. More than 140 of these mutants have been discovered to play a role in familial ALS, even though that represents only about 2 percent of all cases of the disease. A mutant gene creates an abnormal protein, and these aberrant knockoffs of the native SOD protein are obviously crucial to the version of the disease that runs in families. But that's not all: there is growing evidence that even in so-called sporadic cases, which represent the vast majority of ALS, the SOD protein is somehow changed, even if the gene responsible for it has not mutated. It changes shape, perhaps by partially unfolding, and when that happens, this stalwart defender of the cell is not just incapable of defending cellular integrity but becomes destructive itself.

This is something that stumped researchers for some time. How could it be that a misfolded protein, rather than simply depriving the cell of its contribution, actually makes things worse? There have been myriad suggestions, but it looks like aggregations of these mutant SOD proteins are behind it. It might be that the molecular chaperones that help proteins fold properly are so overcommitted trying to fix the misfolded SODs that they're not available to maintain the good health of all the other cellular proteins, and the death of the cell ensues. Or it might be that aggregates somehow reverse the normal function of SOD and *promote*—rather than inhibit—oxidative damage.

The details may not be complete, but a picture of a prion-like process in ALS has now emerged: misfolded proteins are formed and clump together, and somehow cause the death of the cell. Such misfolded proteins are capable both of exiting the cell in which they form and entering neighboring cells. And most recently—and crucially—Dr. Neil Cashman and a group at the Brain Research Centre at the University of British Columbia have shown that certain misfolded versions of SOD will cause the normal version of that protein to misfold. These observations together lay down a molecular foundation for the inexorable progress of the disease: proteins misfolding, recruiting others and then spreading from cell to cell throughout the central nervous system. And there's a chance that treatments for the disease will be found through an understanding of those molecular dynamics.

23

Chronic Traumatic Encephalopathy
The Athletes' Plague

The Lou Gehrig "controversy" described in the previous chapter was triggered by the finding of a very small number of patients in whom blows to the head had apparently led to an ALS-like disease of motor neurons. It diverted the public's attention from the fact that the connection between blows to the head and neurodegenerative disease is becoming stronger practically by the month. The prototype, the aptly named dementia pugilistica, a disease of former boxers colloquially called punch-drunk, was first described in the late 1920s. But today the general (and much less colorful) term for any neurodegeneration caused by repeated blows to the head is CTE: chronic traumatic encephalopathy. CTE has gotten more ink over the past two years than any of the neurodegenerative diseases in the past three chapters, mostly because of its impact on professional athletes.

In particular, professional football players seem unusually prone to CTE, though hockey players are susceptible too. And while in Canada there's been tremendous concern voiced about hits to the head in hockey, they're nothing like in football. A player in the NFL might endure a thousand hits to the head in a single season, many of those in practice. Although repeated concussions are definitely a risk factor, it's not yet clear that a player has to suffer a concussion; instead, he might experience countless subconcussive events, with the same end point.

That end point is the tragic course of CTE. It gradually develops into a dangerous mix of memory loss, outbursts of anger and aggression, loss of self-control, depression and drug addiction.[1] It's not clear how many boxers and professional football and hockey players (and surely MMA—mixed martial arts—fighters also) will eventually develop this disease; what *is* clear is that, once started in the brain, it continues to spread inexorably, even in the absence of further impacts.

The disease features accumulations in the brain of the protein tau, which also is one of the two key diagnostic features of Alzheimer's disease. While it appears that all of us accumulate the twisted pairs of tau filaments in our brains through life, repeated head injuries apparently accelerate that process dramatically.

Dr. Ann McKee was one of the co-authors of the Lou Gehrig paper that resulted from the study referred to in Chapter 22; she is a researcher at the Bedford Veterans Administration Medical Center in Massachusetts and has dissected the brains of several former professional athletes. When I was in her lab, she showed me a stunning set of transverse slices of football players' brains. At the top, slices of Dave Duerson's brain. Duerson was a great, hard-hitting NFL player who went on to become a successful businessman, only to fall victim to CTE and eventually commit suicide at the age of fifty. He shot himself in the chest to ensure that his brain would be intact for autopsy.

His brain was edged with the dark deposits of tau typical of the disease. Below Duerson, Wally Hilgenberg, who was a stand-out guard and linebacker for sixteen years in the NFL, most of those with the Minnesota Vikings. He died when he was sixty-six with what was described as ALS. Hilgenberg's brain has the same dark edging, but in more places, and pushing deeper into the brain tissue. On the bottom row, an unnamed former NFL player who died in his seventies. His brain was the same, only worse. It was easy to see the progression of the disease with advancing age.

Then Dr. McKee showed me the brain of Owen Thomas, a twenty-one-year-old football player at the University of Pennsylvania who had hanged himself. Thomas had played football since he was nine years old and had never officially suffered a concussion but nonetheless had the first traces of the same destruction of tissue seen in the ex–football players' brains. In fact, the areas of damage in Thomas's brain, when viewed under the microscope, were virtually identical to the others.

Hockey players Reggie Fleming, Bob Probert and Derek Boogaard had CTE too. Even though the number of athletes' brains studied is still relatively small—too small to draw any firm conclusions about the incidence of CTE—the results should raise red flags everywhere. There are still many questions: What about all those former athletes who didn't get CTE even though they were subject to the same punishing conditions? What saved them? How many hits does it take? When paired with the previous question, it seems the answer to this one is, it depends who you are. What can be done about it?

The first answer to that is: protect the players. A recent rating of football helmets by Virginia Tech showed that only one got a five-star rating, and others, still in common use in American college football, were found to be significantly less protective when it came to concussive impacts. Football helmets are good at preventing skull fracture but can be practically impotent when it comes to concussion.

The other way to answer that last question is to do exactly what scientists involved in studying CTE are doing: besides examining postmortem brains to collect statistical amounts of data, they are turning to former players still living, scanning their brains yearly, and comparing those images with the results of psychological tests. If these studies start to pin down when the disease process begins, and where, it might be possible to start thinking of some sort of medical intervention while the disease is still in the early stages.

In the meantime, CTE adds another unusual specimen to the mix of neurodegenerative conditions that resemble prion diseases. In this case—remarkably—head trauma triggers the rapid progression of tau accumulation. And again, as we've seen, those misfolded tau spread from neuron to neuron, from one area of the brain to another. Understanding *that* is the key.

The jury is still out on Stan Prusiner's forecast that understanding prion diseases might lead us to revelations about other neurodegenerative conditions. Yes, there are striking similarities, but there's still that one key difference: they are not infectious. Their misfolded proteins seem capable of spreading by multiplying inside the cell and even spreading from cell to cell, even tissue to tissue through the body. But not from one body to another. And that might, in the end, keep the two kinds of disease apart.[2]

Yet it still might happen that some detail of protein misfolding or plaque formation will be recognized as key to all these disease processes, but some crucial questions would have to be answered first. For instance, what exactly do plaques do in these diseases? If you can manage to interrupt their development, will that stall or even reverse disease? And, perhaps most important, what causes them to form in the first place? In the end, the crucial link would be one that led to some sort of treatment for any or all of these conditions.

Speculating even further, a study just released suggests that the "normal" decline in memory with age might have nothing to do with the number of years you've lived and everything to do with the gradual accumulation of Lewy bodies and other agglomerations of misfolded proteins.[3] This study, of some 350 adults, showed that individuals who, on autopsy, had no discernible features of neurodegenerative disease, like accumulations of Lewy bodies or plaques, simply hadn't experienced the typical failure of memory with age. None. Among the others, there were two stages of memory loss, one slow and not normally labeled as "disease," and the other, the precipitous decline seen in Alzheimer's and other diseases. In both kinds, accumulations of misfolded proteins were present—the greater the accumulation, the worse the condition. It's a long shot, but if this process could be slowed or even prevented, the societal benefit would be incalculable. And if insights contributing to a medical triumph of those proportions could be provided by better knowledge of the prion diseases, then Stan Prusiner's bold speculations of thirty years ago would be *very* impressive.

24

And in the End . . .

The prion story is not just about dramatic advances in science, the revelation of a completely new mechanism for infection and the hopes of better understanding and treatments for tragic neurological disease; it has also been a bitter battle between those who believe in the concept of disease by misfolded proteins and those who don't. In the beginning, thirty years ago, when Stanley Prusiner christened prions, anger at and resentment of this PR-savvy and hard-driving scientist were widely shared. But in the years since, as better evidence of prions and their peculiar "protein-only" qualities has been gathered, most of that opposition has faded. But not completely. The flames have died down, but the coals are still glowing. Today, what used to be two rival communities can be reduced (at least symbolically) to two individuals: the ever-dominant Stanley Prusiner on one side and unflagging critic Yale biologist Laura Manuelidis on the other. To read what they write today is to see not just how open to interpretation scientific data can be but how subjective the "objective" world of science can be.

Prusiner first. He has his Nobel Prize, which alone guarantees his place in the history of science. But is that really enough? Nobels are often controversial: Did the recipient actually merit it? Was it granted hastily, before the science was truly settled?

Were other deserving scientists left out? There are always questions and doubts, and so it never hurts to continue to burnish the reputation of even a Nobelist. In Stanley Prusiner's case, there is an article in the *Annual Review of Genetics* that is, in every sense, a celebration of prions and their creator.[1] But first, some background.

The late molecular biologist Gunther Stent wrote extensively on the nature of science, once controversially arguing that the reason some discoveries are not appreciated immediately is that they are simply ahead of their time. His argument is that they, though correct, languish because you can't connect them to the prevailing wisdom; if you're forced to try to make sense of them, they demand a leap of faith that doesn't make sense.

In Stent's opinion, the discovery that the genetic material in living things was DNA, not protein, was a prime example of that. In that case, the team of Avery, McCarty and MacLeod (names that are as familiar among biochemists and geneticists as Watson and Crick) performed experiments in the mid-1940s proving that when pneumonia bacteria transform themselves from being harmless to highly infectious, DNA is the material that makes it happen. A, M and M were building on work done in the 1920s by Frederick Griffith.[2]

Griffith had come up with the original experiments showing that bacteria could change their spots. It took A, M and M to take those original findings and sweat out the deliberate, painstaking, plain hard work to strip away all the pretenders, leaving DNA as the "transforming principle." It wasn't easy, but it was made much more difficult by the fact that no one believed DNA *could* be the genetic material. Avery was very antsy about this. Here's what he wrote his brother Roy in 1943: "But today it takes a lot of well-documented evidence to convince anyone that [DNA], protein free, could possibly be endowed with such biologically active and specific properties and that is the evidence we are now trying to get. It is lots of fun to blow bubbles but it is wiser to prick them yourself before someone else tries to."[3]

What was the hang-up about DNA? Genes were obviously incredibly complex and variable, but DNA was believed to be chemically monotonous. It was impossible to see how you could squeeze any kind of variation out of it. In one way, the scientists who believed this were right: the components of DNA are few in number—four only—and if they were put together in some sort of simply repetitive way, there is no way genes could be made from them. (That's why Stent labeled this discovery as "premature"—you couldn't really make sense of it in its time.) But it isn't mindlessly repetitive, and Watson and Crick, when they figured out the double helical structure of DNA, showed why. There may be only four letters in DNA's alphabet, but they can be written out in endless combinations as they are strung along the axis of the double helix. That's how genes made of DNA can work.

But none of that was known in the 1940s. Proteins, with their twenty amino acids, were the preeminent choice for the genetic material, not DNA. And here were Avery and McCarty (MacLeod had done the early work) with clear evidence that it *was* DNA, not protein. No wonder Avery wanted to keep getting more and more evidence. They finally published their results and, predictably, most scientists didn't believe them. Even though they were right.

Why were their claims rejected by so many? One of the major problems they faced was to eliminate the possibility that there was *any* protein in the material that transformed bacteria. Proving a negative is a touchy business: try to prove there is no such thing as the Loch Ness monster. In this case, opponents could always say, "Sure there's DNA, but have you really eliminated *all* the protein?" Sound familiar? Yes, it is the mirror image of the argument that Stanley Prusiner faced: "Sure there's protein, but have you really eliminated *all* the DNA?" This parallel struck both Prusiner and Maclyn McCarty, one of the famous three, so they sat down and wrote an academic article about it.[4]

The two are old friends, going back to that period in the late 1970s when Prusiner received large amounts of money from a research fund provided by the R. J. Reynolds Tobacco Company. (The case has been convincingly made that the tobacco company was simply trying to establish a reputation for being pro-research—and therefore not hostile to medical evidence pointing to the risks of smoking.) McCarty was one of the administrators of that fund.

Most important is the tone of the article. It is triumphal. It reeks of authority. Prusiner is speaking as the inventor of this entire branch of science, and to underline that position, his derring-do act of scientific heresy is linked to an inarguably classic instance, that of Avery, McCarty and MacLeod, whose evidence also flew in the face of majority opinion.

Michel Morange, a biologist at the École Normale Supérieure in Paris, argues that although there are indeed parallels between these two revolutionary events in biology, there are also some significant differences omitted by the authors.[5] The first is that Avery and colleagues published the first suggestion that DNA might be the genetic material—no one had even considered that before. On the other hand, Prusiner's 1982 touting of a protein had been anticipated by others—especially John Griffith—fifteen years earlier. Morange also claims that the relative impacts of the two species of science are dramatically different: it can be said that the Avery paper revolutionized biology, but it's still too soon to say that Prusiner has done anything remotely similar.

His third point, about style, is the most intriguing. Avery was, as I suggested, extremely reticent to come forward with his results, publishing only when his co-authors started to get impatient. He knew the reaction that would greet them. Prusiner, on the other hand, relished the spotlight, however negative it might be. He was the one who coined a new term for something that was still mysterious; he didn't hesitate to call his idea "heretical"; he described the reaction of scientists to that idea as "dubious," "skeptical," "incredulous," "irate" and even "vicious." There may be doubts about the significance of Stanley Prusiner's contribution to biology, but apparently he doesn't share them.

Nor has he abandoned his efforts to cement the place of the word "prion" in the biological lexicon. In a paper published in June 2012, he refers to the deposits of misfolded proteins in Alzheimer's disease, the plaques called beta-amyloid, as "prions." It's right there in the title of the paper: "Purified and Synthetic Alzheimer's Amyloid Beta (Ab) Prions."[6] As far as I can tell, this is the first time that these have been called "prions," especially notable at a time when other scientists are moving in the opposite direction, lumping all these diseases, prion-based or not, together as "protein misfolding disorders." But Prusiner is pushing back. Scientists whom I have asked about this seem somewhat nonplussed, admitting that "prion-like" would be as far as they would be willing to go.

As I said, most of the doubters have either converted or are in retreat. But there remains a handful who protest that the definitive proof of protein-only infection and reproduction has still not been provided, and the most prominent of these is Yale biologist Laura Manuelidis. She is the last prominent voice of the resistance to Prusiner that has been there from the beginning, but while Prusiner writes from the throne, Manuelidis writes with her back to the wall:

> One may well ask why the scientific community as a whole, in addition to the public at large, believes so totally in the prion model of these infectious agents. I think the casual or expansive use of words, especially those taken out of their properly defined scientific context, has obscured the truth of evidence. . . . It is time to look at the actual molecules that could be part of this structure, rather than making believe that nothing else is needed, and nothing else is there.[7]

Manuelidis has been studying diseases like CJD for years but, from the beginning of the prion era, has steadfastly refused to acknowledge that these diseases are caused by misfolded proteins. As far as she is concerned, prion diseases have all the earmarks of virus infections. Of course, this runs absolutely counter to decades of research, reaching back to the 1960s, which failed to find any evidence of genes associated with the infectious particles, genes that *all* viruses possess. But while Manuelidis would acknowledge that it would be extremely difficult for the pro-prion crowd to establish beyond a shadow of a doubt that prions contain zero genes, she is not putting her eggs in *that* basket: she argues that there is positive evidence of viruses in association with the prion diseases. It would be a mistake to dismiss Manuelidis's reports as complaints and nothing more—she is a creative scientist.[8]

For instance, in 1993, she published a report with her now late husband, Elias, suggesting that a majority of us carry a virus for Creutzfeldt-Jakob disease but that in only very few cases are those viruses activated. A trigger is needed, like ultraviolet light from the winter sun for the cold sore virus. When they inoculated hamsters with white blood cells from normal adults of varying ages, twenty-six of thirty animals came down—eventually—with a spongiform disease. Pretty stunning numbers. To the Manuelidises, it sure looked like the behavior of a virus, albeit one that took a very long time to kick in, and they didn't hesitate saying so.

Laura Manuelidis has tilted toward viruses over prions ever since. She writes irritated book reviews ("Prion biology and disease . . . leads one to re-examine the objectivity of science and whether it is a myth vanished");[9] she claims that "an argument based on the dominant prion theory" delayed recognition of the realization that mad cow disease could infect humans;[10] and she pounces on any inconsistency in prion science: when one study showed that disease in mice sometimes developed in the absence of any detectable prions, Manuelidis argued that while prions might be the result of disease, viruses were the cause.

One of her most recent studies produced photographic evidence of regular arrays of tiny, roughly spherical objects in cells, about twenty-five nanometers (one-millionth of an inch) across, looking like nothing more than assemblies of virus protein coats, neatly packaged inside a membrane, ready to be stuffed with genes and exported from the cell. In the paper she authored in 2007 in the *Proceedings of the National Academy of Sciences,* Manuelidis did some Prusinerizing of her own prose, at first referring to the particles as being "most consistent with a viral structure" but then, a mere five lines later, promoting them to "dense viral spheres."[11] She has repeatedly alluded to the existence of these particles since that publication. But she has a significant challenge: so far, neither she nor anyone else has demonstrated that these particles cause disease. And though the prion theory isn't yet completely airtight, it continues to make strides.

If you stand back a little from Prusiner's and Manuelidis's statements, you are reminded of the argument from social science that people with different preconceptions will not just take a different point of view of the same study but will actually experience a completely *different* study. So where prion scientists interpret a proliferation of different varieties of prion types as a set of slightly differently folded proteins, any one of which might predominate depending on the circumstances, Manuelidis and her colleagues see them as a set of viruses that mutate according to circumstances. Same data, different mindset. Is Manuelidis just clinging to an untenable position, or is it possible that some of her objections point to real gaps in prion theory? I admire her tenacity, but she is increasingly isolated.

But what if you just look at the science itself, or at least that much of the science that is generally agreed upon? Has it been the revolution in thinking that some scientists argue? A revolution is a subjective thing, and the scientific versions are defined differently by different scholars, but the one that stands out is that portrayed by historian of science Thomas Kuhn in his book *The Structure of Scientific Revolutions*.[12] You've already absorbed a part of his book, whether you know it or not, if you've heard the term "paradigm change." Kuhn came up with that, gave it a very specific definition in his book, then likely cringed as it became applied to anything from new models of cars to techniques for moving office furniture around. Everyone in science knows about Kuhn. Most can name a scientific revolution or two, including Alfred Wegener's continental drift, Louis Pasteur's infection theory, Einstein, Darwin. But it isn't only headline science that gets revolutionized. Even out-of-the-way corners of science can be overturned—it's just that in those cases, no one notices. There's no award for having your work rated as being part of—or, even better, triggering—a scientific revolution, just the ego push, which, as we've seen, is definitely part of science.

Kuhn saw things this way: most of the time, any branch of science percolates away, solving the puzzles and mysteries that confront it. During this relatively peaceful but progressive time, scientists share a kind of overarching view of how things work, and that determines the experiments that are done. If you think light is a flood of particles, you won't do experiments that test how it behaves like a wave, or vice versa. The dominant paradigm determines what's true, what's not and what questions can be asked. It enables scientists to focus on detail; it even allows the invention and use of technologies designed to work on the problems posed by the paradigm. At the same time, it is a blinkered view of the world—the odd anomalous result, however curious, is likely to be ignored. That can continue for a long time, but if more and more anomalous results are discovered, tension rises and, sometimes, even though the ruling paradigm can be tweaked and stretched to some degree to accommodate these new results, it may eventually reach a breaking point.

Even then it's possible that nothing will change. If scientists simply can't make sense of the paradigm, or connect the dots from the peculiar observations to it, then those observations are likely to be shelved. (Of course, this is what most proponents of everything from Bigfoot to ESP think has happened in their field—nothing wrong with the observations, but scientists are simply too straitjacketed in their worldview to accommodate such claims.) But every once in a while, the anomalies are just too much and the whole paradigm is thrown into question. A period of chaos ensues, where both the old and the new paradigms compete to explain both the old and the new observations.

Usually by the time it reaches this stage, the former paradigm is overthrown and the new one rules—it's a paradigm shift. The new one doesn't even have to be correct; it just has to explain things a little better. But as Kuhn pointed out, sometimes this isn't exactly an orderly process. It can happen—and certainly did in the early days of the prion controversy—that scientists on opposite sides of the fence talk right through each other by choosing different data to argue for their side and ignoring annoying data that might contradict it. Stan Prusiner emphasized the evidence for a protein and against genes; his opponents stressed that there were strains that could be best explained by the existence of genes, not protein. As Kuhn wrote about most scientists in this situation, "Both are looking at the world. . . . But in some areas they see different things."[13]

So, are we looking at a revolution? One thing is for sure: Thomas Kuhn's own description of how a scientific revolution takes hold is eerily reminiscent of the prion story:

> At the start a new candidate for paradigm may have few supporters, and on occasions the supporters' motives may be suspect. Nevertheless, if they are competent, they will improve it, explore its possibilities, and show what it would be like to belong to the community guided by it. And as that goes on, if the paradigm is one destined to win its fight, the number and strength of the persuasive arguments in its favor will increase. More scientists will then be converted, and the exploration of the new paradigm will go on.[14]

Later in that passage, Kuhn says that the paradigm will roll on "until at last only a few elderly holdouts remain."[15] In this he is echoing the renowned physicist Max Planck, who famously claimed that new scientific truths don't make people see the light, but rather the opponents eventually die out, and the new generation just grows up with the new truths, with which, of course, they will one day become obsessed themselves.

And that's where things stand right now. In one way, you might see the prion story as an example of how the wheels of science grind slowly. It's been thirty years since Stan Prusiner aroused hostility and praise in roughly equal amounts with his christening of prions. In that sense, considering there are still some scientists who don't believe in them, progress has been incremental. On the other hand, prions have revolutionized thinking about how disease can spread, they have alerted us to the possibility of new and unanticipated outbreaks and they still may open doors to the understanding of conditions like Alzheimer's and Parkinson's. And it's all been a dramatic—often tragic—tale peopled by an amazing cast of characters. For that alone it is worth telling.

If you stand back from the day to day of the prion scientific revolution, you get the impression that the most dramatic moments, the fiercest thrusts and parries, are over; the blood has been mopped up; and we're now in a quieter stage that might correspond to what Thomas Kuhn characterized as puzzle solving, a time when scientists who grew up with the new paradigm set out to clarify it, to track down and solve the anomalies. It may seem like that's where we *should* be, but it doesn't feel like it. Mysteries still abound, some fundamental pieces have not found their place in the puzzle and this story is far from over.

Acknowledgments

This book has had a lengthy gestation, and if I thanked everyone who added some bit of pertinent advice along the way, this would be one unwieldy piece of writing. So with apologies to those who likely did contribute but will go unmentioned, here goes:

First a big thank-you to my friends and colleagues in the prion world who not only answered my questions along the way but also betrayed no impatience or skepticism when I repeatedly assured them that this book would one day be published. They include Neil Cashman, David Westaway, Robert Will, Mike Miller, Avi Chakrabartty, Aru Balachandran, the PrioNet board of directors and especially Val Sim and Stefanie Czub. Stefanie showed me around her lab in Lethbridge and let me reexperience that chilling moment when she confirmed the first Canadian case of BSE; Val patiently filled in the gaps in my knowledge.

Sometimes the best way to clarify your thoughts is to be obliged to say them out loud, and I had several wonderful opportunities to do that, helped immeasurably by Lyanne Foster at the Alberta Prion Research Institute, Sandy Haney at PrioNet, Stephen Moore and Kevin Keough.

In the same way, visuals, whether still or animated, help immeasurably in choosing the right words. Lin Tang drew the illustrations that accompany the text in this book, and I couldn't have hoped for anything better. Both Alex Tirabasso and Vikram Mulligan created stunning animations, which I have used in talks, and I can say without any fear of contradiction that there is nothing else anywhere like them. You can't see them here, but they informed everything I wrote.

One of the things that we emphasize at the Banff Science Communications Workshop every summer is the importance of editing. Without it, authors are simply not as good. Jim Gifford and Alex Schultz split the duties in this case, and the book is much better for that.

Jackie Kaiser, my agent, has been a strong supporter through the whole writing process, and I'd bet she is delighted with the book, not least because it's actually done.

Mary Anne actually claimed all along that she wanted to read the manuscript, so you can't get much better support than that, and I always have Rachel, Amelia and Max in mind when I try to write clearly and vividly.

Notes

Introduction

1. Newton in a letter to Hooke, February 5, 1676, in *The Correspondence of Isaac Newton*, ed. H. W. Turnbull, J. F. Scott and A. R. Hall (Cambridge: Cambridge University Press, 1959–1977), vol. I, 416.

2. Newton definitely ranks as one of the most recognizable scientists, but mostly for the idea that he saw the apple fall . . . or, as many think, was *hit on the head* with the apple. But there is absolutely no evidence it happened that way. He saw an apple fall, period.

3. Marshall was spared the worst of the effects of *H. pylori* infection because his case cured itself spontaneously, but a colleague who ingested the bacteria wasn't so lucky. It took three years and endless rounds of antibiotics before he was free of the bacteria—and the symptoms.

1. The Mystery of Kuru

1. Gajdusek to Joseph Smadel, March 15, 1957, in *Kuru Early Letters and Field-Notes from the Collection of D. Carleton Gajdusek*, ed. Judith Farquhar and D. Carleton Gajdusek (New York: Raven, 1981), 8.

2. Gajdusek to Vin Zigas and Jack Baker, August 29, 1957, in ibid., 126.

3. Gajdusek to Jack Baker, August 30, 1957, in ibid., 129.

4. Sir Macfarlane Burnet to John Gunther, in ibid., 41.

5. Gajdusek to Joseph Smadel, September 18, 1957, in ibid., 159.

6. Leonard Kurland to Henry Imus, August 20, 1957, in ibid., 117. It turned out that pellagra was caused by consuming corn that had not been prepared properly; wage-earning men had better access to a variety of food, women gave their children the best of the rest, and even higher estrogen levels might have accounted for the gender disparity in the disease.

7. Burnet to A. L. G. Rees, December 6, 1957, in ibid., 264.

2. Barflies and Flatworms

1. If you want something a little more lurid that this, try the first chapter of Richard Rhodes's *Deadly Feasts* (New York: Simon and Schuster, 1997).

2. Gajdusek to Joseph Smadel, November 17, 1957, in *Kuru Early Letters and Field-Notes from the Collection of D. Carleton Gajdusek,* ed. Judith Farquhar and D. Carleton Gajdusek (New York: Raven 1981), 255.

3. In an odd example of rewriting history, Gajdusek first described having had these ideas at "one rather jocular stage of hypothesizing," but in a later account changed that to "one rather *casual* stage of hypothesizing." I can only guess being "jocular" about kuru could easily be misinterpreted and, if so, misleading, because it is clear from everything Gajdusek wrote that he was respectful and empathic toward the Fore.

4. In fact, in a recent survey, scientists could find only one case in the entire animal world of infection spread by cannibalism. On the Canary Islands, lizards spread a single-celled parasite among themselves by using a peculiar analogue of group cannibalism. An infected lizard, when attacked by another, sheds its tail to distract the aggressor and runs for its life. The attacker consumes the tail, the escapee grows another and the whole cycle is repeated. One parasite-laden lizard can infect as many others as tails it can regrow. But this odd case aside, cannibalism is nearly negligible as an effective way of spreading disease.

5. The phrase "born after the ban," while entirely appropriate here, was coined for another, related and much better-known disease: BSE, or mad cow disease.

6. Hank Nelson, "Kuru: The Pursuit of the Prize and the Cure," *Journal of Pacific History* 30(2) (1996): 196.

7. For more details on McConnell see "Consumed by Learning" in my book *The Barmaid's Brain and Other Strange Tales from Science* (Toronto: Penguin Canada, 1999).

8. R. Glasse, "Cannibalism in the Kuru Region of New Guinea," *Transactions of the New York Academy of Sciences* 29(6) (1967): 748–54.

3. Cannibalism

1. In *Kuru Early Letters and Field-Notes from the Collection of D. Carleton Gajdusek*, ed. Judith Farquhar and D. Carleton Gajdusek (New York: Raven Press, 1981), xxiii.

2. William Arens, "Rethinking Anthropophagy," in *Cannibalism and the Colonial World*, ed. Francis Barker, Peter Hulme and Margaret Iversen (Cambridge: Cambridge University Press, 1998), 50–51. Of course, in the article referred to, Gajdusek was still arguing that kuru was transmitted not by cannibalism but rather by indirect infection of hands and eyes.

3. Michael Harner, "The Ecological Basis for Aztec Sacrifice," *American Ethnologist* 4(1) (1977): 117–35.

4. Annette Beasley, "Kuru Truths: Obtaining Fore Narratives," *Field Methods* 18(1) (2006): 33.

5. Vincent Zigas, *Laughing Death: The Untold Story of Kuru* (New York: Humana Press, 1990), 220.

6. Lyle Steadman and Charles Merbs, review of *Kuru: Early Letters and Field-Notes from the Collection of D. Carleton Gajdusek*, ed. Judith Farquhar and D. Carleton Gajdusek, *American Anthropologist* 84(3) (1982): 611–27.

4. Igor and Bill

1. Igor Klatzo to Gajdusek, September 13, 1957, *Kuru Early Letters and Field-Notes from the Collection of D. Carleton Gajdusek*, ed. Judith Farquhar and D. Carleton Gajdusek (New York: Raven Press, 1981), 155.

2. Although there is an intriguing suggestion that the origin of a Chinese character for "itchy" might be a combination of two ancient characters, meaning "disease" and "sheep." These ancient characters are likely more than fifteen hundred years old, suggesting that scrapie may have been present in China that long ago.

3. The flying shuttle meant that one weaver, rather than two, could manage to throw the weaving shuttle from one side to the other of the loom, meaning either an instant halving of the number of weavers or a

doubling of their output, the kind of advance on which the Industrial Revolution thrived.

4. W. Hadlow, "The Scrapie-Kuru Connection: Recollections of How It Came About," in *Prion Diseases of Humans and Animals,* ed. Stanley Prusiner et al., 40–45 (Chichester, UK: Ellis Horwood, 1993), 40.

5. W. Hadlow, "Scrapie and Kuru," *The Lancet,* September 5, 1959: 289–90.

6. Hadlow, "The Scrapie-Kuru Connection," 43.

7. Ibid., 45.

5. The Life of a Cell

1. Newton, you should know, is said to have laughed only once in his life. What triggered this outburst of mirth? Someone to whom he had lent a copy of Euclid asked the great man what use its study would be to anyone. As far as Newton was concerned, this was a real knee-slapper.

2. L. L. Larison Cudmore, *The Centre of Life: A Natural History of the Cell* (Newton Abbot, UK: David and Charles, 1978), 6.

3. The only memorization task worse? Meiosis.

6. The Death of a Cell

1. David Morens and Anthony Fauci, "The 1918 Influenza Pandemic: Insights for the 21st Century," *Journal of Infectious Diseases* 195 (2007): 1018–28. This passage continues by expanding on the storm analogy: "A deleterious overexuberant release of proinflammatory cytokines" (p. 1022), a phrase I initially read as "a *delirious* overexuberant release." I actually think that's an improvement.

7. When Is a Virus Not a Virus?

1. Rabi's comment ranks right up there with Wolfgang Pauli's acid denunciation of an idea of physics he thought worthless: "That's not right. It's not even *wrong.*"

2. William Hadlow, "Reflections on the Transmissible Spongiform Encephalopathies," *Veterinary Pathology* 36 (1999): 524.

3. Allan Dickinson, 1999 interview with Kiheung Kim, in Kiheung Kim, *The Social Construction of Disease* (New York: Routledge, 2007), 22.

4. J. T. Stamp et al., "Further Studies on Scrapie," *Journal of Comparative Pathology* 69 (1959): 268–80.

5. When Alper was seventy-nine, a burglar appeared at her front door, armed and wearing a balaclava. She assaulted him, grabbing at his gun. He ran away in a panic.

6. T. Alper et al., "The Exceptionally Small Size of the Scrapie Agent," *Biochemical and Biophysical Research Communications* 22(3) (1966): 283.

7. I don't really buy the idea that the journal wasn't first-rate, but if you dig it out of the library, as I did, you can't help but be struck by the fact that it looks like someone simply bound together a bunch of articles that they typed themselves. Double-spaced.

8. Francis Crick, "On Protein Synthesis," in *Symposia of the Society for Experimental Biology* 12, (1958): 153.

9. Francis Crick, "Central Dogma of Molecular Biology," *Nature* 227 (1970): 562.

10. There are some exceptions, though. David Bolton, a scientist who has worked on these diseases for years, wrote that Griffith's idea "provided the foundation for all subsequent modern versions of the protein hypothesis, although that fact is often not recognized. Griffith published just this one paper in this field but it forever changed the way the world was viewed." David Bolton, "Prions, the Protein Hypothesis, and Scientific Revolutions," in *Prions and Mad Cow Disease,* ed. Brian Nunnally and Ira Krull (New York: Marcel Dekker, 2003), 34. Bolton has a role in a long-standing anecdote about Griffith's paper; that is, that he discovered the Griffith paper only because he was photocopying the article just before it in that issue of *Nature.* The last column of that article and the first of Griffith's are on the same page.

11. The name is apparently a kind of homage to the neutrino, the "little neutron." A calculated homage, in that a good name helps get attention if nothing else.

8. Creutzfeldt-Jakob Disease

1. C. Bernoulli et al., "Danger of Accidental Person-to-Person Transmission of Creutzfeldt-Jakob Disease by Surgery," *The Lancet,* April 26, 1977: 478–79. The patients described in this paper died in 1976. Just too late, a report was published in that same year showing that formaldehyde does not inactivate the CJD agent.

2. In Richard Rhodes, *Deadly Feasts* (New York: Simon and Schuster, 1997), 143.

3. One of the relatively unsung advances of genetic engineering was the invention of a way of synthesizing human growth hormone in bacteria. Sadly, the final approval of this product came too late for the many victims, but at least now patients don't have to choose between turning down treatment or risking CJD.

4. Esther Kahana et al., "Creutzfeldt-Jakob Disease: Focus among Libyan Jews in Israel," *Science* 183 (1974): 91.

5. L. Herzberg et al., "Creutzfeldt-Jakob Disease: Hypothesis for High Incidence in Libyan Jews in Israel," *Science* 186 (1974): 848.

9. Magnificent Molecules

1. While it is certainly true that a single amino acid substitution can be deadly, it's also true that most proteins can sustain a number of changes. Those changes are ones that only alter the shape slightly, not enough to affect the workings of the protein. This is true even of hemoglobin.

10. Protein Origami

1. http://www.psc.edu/science/kollman98.html.

11. Stanley Prusiner's Heresy

1. The birds are now relegated to the sixteenth page of a Google search: prions as Procellariiformes, in Britannica Online, and while

the definition was correct, the six supporting online articles were, you guessed it, about the protein. Some of Dr. Nevitt's colleagues have wryly suggested that they wish Prusiner had called his invention "stanons."

2. Stanley B. Prusiner, "Novel Proteinaceous Infectious Particles Cause Scrapie," *Science* 216 (1982): 136–44.

3. Gary Taubes, "The Name of the Game Is Fame: But Is It Science?" *Discover,* December 1986, 28. There are much worse names too; in a 1991 article, D. H. Adams suggested "scaies"—small cell-associated infectious entities. D. H. Adams, "Does the Infective Agent of Scrapie Replicate without Nucleic Acid? An Assessment," *Medical Hypotheses* 35 (1991): 253–64.

4. Prusiner, "Novel Proteinaceous Infectious Particles," 141.

12. An Infectious Idea

1. I still have his e-mail rejection of a letter I wrote requesting an interview. Well, it wasn't exactly "his"; it came from a member of his lab.

2. Gary Taubes, "The Name of the Game Is Fame: But Is It Science?" *Discover,* December 1986, 28–52.

3. Ibid., 33.

4. Anon., *The Lancet,* May 29, 1982: 1222. The fact that Stanley was wrong didn't prevent his getting the Nobel Prize for his work in 1946, long before the understanding of the role of genes in virus replication.

5. D. C. Gajdusek, "Subacute Spongiform Virus Encephalopathies Caused by Unconventional Viruses," in *Subviral Pathogens of Plants and Animals: Viroids and Prions,* ed. K. Maramorosch and J. J. McKelvey Jr., 483–544 (New York: Academic Press, 1985), 493.

6. One thing can't be denied: Prusiner was a force. He marshaled all the help he could get to move prion science forward, including accepting funding from the R. J. Reynolds Tobacco Company, the makers of many different brands of cigarettes, including Camel and Salem. The company funded a variety of medically related topics, including, in this

case, prions. It's been pointed out that this apparent commitment to science made it easier for the company to use scientific arguments that smoking wasn't a proven cause of cancer.

7. One highly regarded prion researcher, having experienced the same situation as Patricia Merz, likes to call it "pre-confirming" Prusiner.

8. Carol Reeves, "An Orthodox Heresy," *Science Communication* 24(1) (2002): 98–122.

9. Laura Manuelidis, "The Force of Prions," *The Lancet* 355 (June 10, 2000): 2083.

10. Stanley B. Prusiner, "Autobiography," http://nobelprize.org/nobel_prizes/medicine/laureates/1997/prusiner-autobio.html.

11. Stanley B. Prusiner, "Prions," *Scientific American* 251(4) (1984): 50.

12. Stanley B. Prusiner, "The Prion Diseases," *Scientific American* 272(1) (1995): 48.

13. A Portrait of the Prion

1. F. J. Donahoe, "Anomalous Water," *Nature* 224(198): 1969.

2. Stanley B. Prusiner, "The Prion Diseases," *Scientific American* 272(1) (1995): 51. Note that Prusiner was admitting the possibility of a mistake only years after it was shown that he was right.

3. Although, to be fair, those who still believe the infectious agent could be some sort of virus argue that removing normal prions from an animal is in effect removing proteins that serve as a portal, or receptor, for the agent (whatever it is), allowing it to gain access, so that this evidence could be cited as support for either theory.

4. Deer, sheep, camels, rabbits, bats, pigeons, quail, ducks, turtles, trout, carp, frogs and more.

5. J. R. Criado et al., "Mice Devoid of Prion Protein Have Cognitive Deficits That Are Rescued by Reconstitution of PrP in Neurons," *Neurobiology of Disease* 19 (2005): 255–65.

6. A. Papassotiropoulos et al., "The Prion Gene Is Associated with

Human Long-Term Memory," *Human Molecular Genetics* 14(15) (2005): 2241–46.

14. Mad Cow Disease

1. Committee of the BSE Inquiry, *The BSE Inquiry: The Report,* vol. 3, *The Early Years, 1986–88* (British Parliament).

2. Ibid., 32.

3. Ibid.

4. Ibid., 33.

5. Ibid., 40.

6. Ibid., 50.

7. Why couldn't it simply jump directly from sheep to cows? Because 20 percent of the cases came from farms with no sheep. Wilesmith considered other possibilities too: the epidemiological team ruled out herbicides, pesticides, vaccines, hormones and antiparasite medications as possible causes, largely because none of these was connected to every case.

8. In one survey of thirteen veterinarians, eleven thought they had seen BSE before 1985, but hadn't thought much about it because it resembled other conditions.

15. Mad Cow in Humans

1. T. A. Holt and J. Phillips, "Bovine Spongiform Encephalopathy," *British Medical Journal* 296 (1988): 1581–82.

2. Ibid., 1582.

3. Anon., "The Man Who Dared to Foretell the Future," *British Medical Journal* 316 (1998): 1190.

4. H. F. Baker and R. M. Ridley, "Bovine Spongiform Encephalopathy," *British Medical Journal* 297 (1988): 133.

5. Committee of the BSE Inquiry, *The BSE Inquiry: The Report,* vol. 6, *Human Health* (British Parliament), 377.

6. As this is written, but one or two are added every year.

7. One young victim, Clare Tompkins, died at the age of twenty-four in 1997, twelve years after becoming a vegetarian. Although there was a suggestion that she had inhaled BSE prions with tiny airborne fragments of dog food from where she worked (which would be a unique mode of transmission for prions), it is quite possible that she consumed contaminated beef products sometime in the mid-1980s.

8. And what could that protection possibly be? You'd think that maybe having two versions of the protein might make it more difficult to build up the momentum of misfolding and aggregation that underlies disease, but nobody really knows at this point.

9. In fact, only about 10 percent of all CJD victims are heterozygotes, even though they make up 50 percent of the population.

10. J. Collinge et al., "Kuru in the 21st Century: An Acquired Human Prion Disease with Very Long Incubation Periods," *The Lancet* 367 (2006): 2068.

11. M. Soldevila et al., "The Prion Protein Gene in Humans Revisited: Lessons from a Worldwide Resequencing Study," *Genome Research* 16 (2006): 231–39.

12. Or has it? As I was writing this, it was announced that the first case of a valine/valine vCJD patient had been discovered in England. Obviously, a huge amount of attention has been directed to this patient, a thirty-nine-year-old woman, because she could be the index case of the second wave of vCJD. But her case was a little unusual in that although she did exhibit some of the brain damage typical of vCJD, her prions were subtly different, enough that the researchers involved were being very careful to point out that this might be some sort of anomalous case—it's simply too soon to tell.

13. There has since been another report of transfusion-caused disease. Also, the one individual who had been infected but died of an aneurysm was an m/v heterozygote. Would he have eventually developed the disease or would his genetics have protected him? We'll never know.

14. P. Clarke et al., "Is There the Potential for an Epidemic of Variant

Creutzfeldt-Jakob Disease via Blood Transfusion in the UK?," *Journal of the Royal Society Interface* 4(15) (August 2007): 675–84.

15. Although an even bigger study of one hundred thousand appendixes/tonsils is under way.

16. More recent figures show four cases of vCJD prions in nearly fourteen thousand appendixes.

16. The Americas

1. R. F. Marsh and G. R. Hartsough, "Evidence That Transmissible Mink Encephalopathy Results from Feeding Infected Cattle," in *Proceedings of the IV International Congress on Fur Animal Production,* ed. B. D. Murphy and D. B. Hunter, 204–7 (Toronto: Canada Mink Breeders Association, 1988).

2. W. J. Hadlow and Lars Karstad, "Transmissible Encephalopathy of Mink in Ontario," *Canadian Veterinary Journal* 9(8) (1968): 193–96.

3. APHIS Fact Sheet, Veterinary Services, United States Department of Agriculture, February 2002.

4. An ongoing puzzle is how the United States has managed to identify only *two* indigenous cases of BSE (with an additional one that was born in Canada), given a cattle population much higher than the Canadian counterpart. Both U.S. cases were an atypical form of BSE found in older cows. There is no convincing explanation for the rarity of BSE in the United States.

17. Into the Wild

1. A. Oyer et al., "Long Distance Movement of a White-Tailed Deer Away from a Chronic Wasting Disease Area," *Journal of Wildlife Management* 71(5) (2007): 1635–38.

2. M. W. Miller et al., "Lions and Prions and Deer Demise," *PLoS One* 3(12) (2008): e4019.

3. It's easier to explain susceptibility to predators in the later stages of the disease, when alertness is compromised. And as far as the

mountain lions are concerned, there has not been—to date—any record of a mountain lion dying from CWD, even though they are, as a species, susceptible to prion disease.

4. R. C. Angers et al., "Prions in Skeletal Muscles of Deer with Chronic Wasting Disease," *Science* 311 (2006): 1117.

5. As of this writing, no signs of disease have yet appeared in any of the people who attended the feast. Ralph Garruto, a biomedical anthropologist who's monitoring the group, told me that they would extend the surveillance for another year, then stop the official study but continue to maintain contact with all of them.

6. M. A. Barria et al., "Generation of a New Form of Human PrP inVitro by Inter-Species Transmission from Cervid Prions," *Journal of Biological Chemistry* 286 (March 4, 2011): 7490–95.

7. Although it might seem incredibly far-fetched or trivial, wild bison also live within the range of CWD already, especially in southwestern Saskatchewan. Obviously, they don't represent huge numbers, but it is known that bison are susceptible to BSE. One researcher told me that the BSE-infected bison brain she examined was one of the most "intense" cases she'd seen. But remember, that was infection by BSE, not CWD.

8. Gordon B. Mitchell et al., "Experimental Oral Transmission of Chronic Wasting Disease to Reindeer (Rangifer tarandus tarandus)" *PLoS One* 7(6): e39055. doi:10.1371/journal.pone.0039055

18. Origins

1. Although it is true that the very same genetic susceptibility that some have to kuru and vCJD, the famous site 129 on the prion molecule, also plays a role in determining one's likelihood of getting CJD. There's something special about that particular place in the molecule.

2. More food for thought (and speculation) was provided by an experiment showing that when the atypical BSE called BASE was injected into a monkey, a cynomolgus macaque, it developed symptoms that were strikingly similar to those exhibited by people who come

down with a rare form of CJD called MM2 (which makes up about 4 percent of all cases).

3. J. Berger, E. Weisman and B. Weisman, "Creutzfeldt-Jakob Disease and Eating Squirrel Brains," *The Lancet* 350 (1997): 642.

4. For the culinarily curious: 8 squirrel brains (scrambled), 1 lb. venison, 1/2 gallon chicken broth, 1 1/2 cups sautéed peppers and onions, 3/4 cup potatoes and a pinch of salt. Mix in pot and let simmer for three hours. In east Texas, squirrel brains have been eaten as dumplings in soup. There was a case of CJD there too.

5. Alan and Nancy Colchester, "The Origin of Bovine Spongiform Encephalopathy: The Human Prion Disease Hypothesis," *The Lancet* 366 (2005): 856–61.

6. http://www.cbc.ca/news/health/story/2005/09/01/new_BSe_theory20050901.html.

19. Cats but Not Dogs

1. Strains were so challenging conceptually for the idea of protein misfolding that Stan Prusiner once declared that he didn't believe in them.

2. Jue Yuan et al., "Insoluble Aggregates and Protease-Resistant Conformers of Prion Protein in Uninfected Human Brains," *Journal of Biological Chemistry* 281(46) (2006): 34848, 34855, 34856.

3. By contrast, recent research shows that the kuru and sporadic CJD prions behave very similarly in mice, reinforcing the idea that kuru likely began when a single individual who had come down with CJD was consumed at a funerary feast.

4. E. Asante et al., "BSE Prions Propagate as Either Variant CJD-Like or Sporadic CJD-Like Prion Strains in Transgenic Mice Expressing Human Prion Protein," *EMBO Journal* 21(23) (2002): 6358–66.

5. R. Barron et al., "High Titers of Transmissible Spongiform Encephalopathy Infectivity Associated with Extremely Low levels of PrP[sc] in Vivo," *Journal of Biological Chemistry* 282(49) (2007): 35878–79.

6. G. R. Mallucci et al., "Targeting Cellular Prion Protein Reverses

Early Cognitive Deficits and Neurophysiological Dysfunction I Prion-Infected Mice," *Neuron* 53 (2007): 325–35.

20. Alzheimer's Disease

1. Stanley B. Prusiner, "Novel Proteinaceous Infectious Particles Cause Scrapie," *Science* 216 (1982): 143.

2. M. Waldman and M. Lamb, *Dying for a Hamburger* (New York: St. Martin's Press, 2004).

3. Ibid., 156.

4. David Snowdon, "Aging and Alzheimer's Disease: Lessons from the Nun Study," *Gerontologist* 37(2) (1997): 150–56.

5. Just recently there has been some direct evidence that people who have plaques and tangles in their brains, but who are cognitively normal like Sister Mary, show evidence of ramped-up activity in those neurons that would be most affected. They are bigger, they could be growing new extensions and they appear to be fighting back, either repairing the damage caused by the accumulating plaques and tangles, or rerouting nerve circuitry to adapt to damage already done. J. C. Troncoso et al., "The Nun Study," *Neurology* 73(9) (2009): 665–73.

6. Here's an example: "I was born in Eau Claire Wisc. on May 24th, 1913, and was baptized in St. James Church." There are seven ideas nested together in this sentence, including, "I was born," "born in Eau Claire Wisc.," "born on May 24th, 1913" and so on. That is 7 ideas in 18 words, or 3.9 ideas in 10 words, which means an idea density score of 3.9. That's in the ballpark of the average score for those who went on to develop Alzheimer's disease, and that's indeed what happened to this sister (even though she had a master's degree).

By contrast, the following passage is much more idea dense, and the sister who wrote it did not suffer dementia: "The happiest day of my life so far was my First Communion Day which was in June nineteen hundred and twenty when I was but eight years of age, and four years later in the same month I was confirmed by Bishop D. D. Mc-Gavick. In nineteen hundred and twenty-six I was graduated from the eighth grade and now my great desire of entering the convent was soon to be gratified." Her idea density score was 8.6, a high score that would predict exactly what happened to her. Those nuns who went on to develop Alzheimer's scored an average of 3.86; those who didn't, 4.78. Both groups declined with age, but if you start higher, you end up higher.

The outstanding question is, do high linguistic scores at the age of twenty have something to do with the capacity for brain reserve shown many decades later? Unfortunately, the Nun Study (which detected bigger neurons in the brains of those who were cognitively intact despite having plaques and tangles) could not confirm that link: there were too few patients whose essays had been kept.

Quotations above from D. Snowdon et al., "Linguistic Ability in Early Life and Cognitive Function and Alzheimer's Disease in Late Life," *Journal of the American Medical Association* 275(7) (1996): 528–32; 530.

7. Oddly enough, there is also lab evidence that the other hallmark of Alzheimer's, the tangles found inside brain cells, can spread, prion-like, through the brains of mice, just as prions would. But being inoculated with the material isn't the same as "catching" it. All this shows is that under certain contrived conditions, proteins involved in Alzheimer's can behave like prions. That doesn't mean they *are* prions.

21. Parkinson's Disease

1. It's harder to explain how head injuries predispose people to Parkinson's, but they do. Research at the Mayo Clinic found that suffering a head injury raises the risk of Parkinson's fourfold; the risk was much higher if the injury was serious enough to require hospitalization or if the patient lost consciousness. Such injuries might kill brain cells or cause abnormal proteins to be released. No one really knows, but regardless, the disease would still not appear for decades.

2. H. Widner et al., "Bilateral Fetal Mesencephalic Grafting in Two Patients with Parkinsonism Induced by 1-Methyl-4-Phenyl-L,2,3,6-

Tetrahydropyridine (MPTP)," *New England Journal of Medicine* 327 (1992): 1556–63.

3. J. Li et al., "Lewy Bodies in Grafted Neurons in Subjects with Parkinson's Disease Suggest Host-to-Graft Disease Propagation," *Nature Medicine* 14(5) (2008): 501–3.

4. F. Pan-Montojo et al., "Progression of Parkinson's Disease Pathology Is Reproduced by Intragastric Administration of Rotenone in Mice," *PLoS One* 5(1) (2010): e8762.

22. Lou Gehrig's Disease

1. A. C. McKee et al., "TDP-43 Proteinopathy and Motor Neuron Disease in Chronic Traumatic Encephalopathy," *Journal of Neuropathology and Experimental Neurology* 69(9) (2010): 918–29.

23. Chronic Traumatic Encephalopathy

1. It hasn't escaped the notice of CTE researchers that a well-known former football player, O. J. Simpson, has shown the characteristic signs of anger management issues and lack of self-control. No word on whether he plans to donate his brain to science.

2. There is one exception: a rare group of diseases called the amyloidoses.

3. R. S. Wilson et al., "Neurodegenerative Basis of Age-Related Cognitive Decline," *Neurology* 75(12) (2010): 1070–78.

24. And in the End . . .

1. Stanley B. Prusiner and Maclyn McCarty, "Discovering DNA Encodes Heredity and Prions Are Infectious Proteins," *Annual Review of Genetics* 40 (2006): 25–45.

2. Name sound familiar? It should: Fred was John Griffith's uncle. John was the mathematician and biophysicist who showed back in 1967 that proteins could multiply without using DNA or RNA. Obviously, insightful science ran in the Griffith family.

3. Oswald Avery to Roy Avery, May 13, 1943, Profiles in Science, National Library of Medicine, http://profiles.nlm.nih.gov/CC/B/D/B/F/_/ccbdbf.pdf.

4. Prusiner and McCarty, "Discovering DNA Encodes Heredity."

5. Michel Morange, "What History Tells Us VIII: The Progressive Construction of a Mechanism for Prion Diseases," *Journal of Biosciences* 32(2) (2007): 223–27.

6. Jan Stohr et al., "Purified and Synthetic Alzheimer's Amyloid Beta (Ab) Prions," www.pnas.org/cgi/doi/10.1073/pnas. PNAS early edition 1–6.

7. Laura Manuelidis, "Transmissible Encephalopathies: Speculations and Realities," *Viral Immunology* 16(2) (2003): 136.

8. And artist. Her first degree was in poetry, she has been a poet ever since and no less than the Russian poet Yevgeny Yevtushenko has praised her writing: "She has a rare quality: her passion is subtle." Introductory essay in Laura Manuelidis, *Out of Order* (Lincoln, NE: iUniverse, 2007), xi.

9. Laura Manuelidis, "The Force of Prions," *The Lancet* 355 (2000): 2083.

10. Laura Manuelidis, "Penny Wise Pound Foolish—A Retrospective," *Science* 290 (5500) (Letters, December 22, 2000): 2257.

11. Laura Manuelidis, "Cells Infected with Scrapie and Creutzfeldt-Jakob Disease Agents Produce Intracellular 25-nm Virus-Like Particles," *Proceedings of the National Academy of Sciences* 104 (2007): 1968.

12. Thomas Kuhn, *The Structure of Scientific Revolutions* (Chicago: University of Chicago Press, 1970).

13. Ibid., 150.

14. Ibid., 159.

15. Ibid.

Index